Rocks and

Minerals

OF THE WESTERN

UNITED STATES

Other books by Elsie Hanauer:

Bits of Knowledge
Handbook of Carving and Whittling
A Handbook of Crafts
Horse Owner's Concise Guide
The Old West: People and Places
Dolls of the Indians
Creating with Leather
The Art of Whittling and Woodcarving
Guns of the Wild West
How to Make Egg Carton Figures
The Science of Equine Feeding
Disorders of the Horse

ROCKS AND
MINERALS
OF THE WESTERN
UNITED STATES

By Elsie Hanauer

South Brunswick
New York: A. S. Barnes and Company
London: Thomas Yoseloff Ltd

A. S. Barnes and Co., Inc.
Cranbury, New Jersey 08512

Thomas Yoseloff Ltd
108 New Bond Street
London W1Y OQX, England

Library of Congress Cataloging in Publication Data
Hanauer, Elsie V
  Rocks and minerals of the Western United States.

  Bibliography: P.
Rocks—Collectors and collecting—The West. 2. Mineralogy—Collectors
and collecting—The West.
I.   Title.
QE445.W47H36   549'.075   73-144
ISBN 0-498-01273-5

Printed in the United States of America

Contents

Introduction 7

The Formation Of Rocks & Minerals 11

The Six Crystal Systems 14

Minerals According To Hardness 15

Becoming Familiar With Rocks And Minerals 21

Descriptions Of Rocks & Minerals 24

Tools and Equipment 31

Where To Find Specimens 33

Safety In The Field 34

Specimen Sizes To Collect 35

Protecting Specimens 36

Field Notes 37

Cleaning Specimens 38

Housing and Display 39

Gem And Mineral Locations 41

   Collecting Localities 43

      Arizona 45

      California 53

      Colorado 69

      Idaho 75

      Kansas 81

      Montana 91

      Nebraska 95

      Nevada 99

      New Mexico 107

      North Dakota 117

      Oklahoma 123

      Oregon 131

      South Dakota 141

      Texas 152

      Utah 157

      Washington 163

      Wyoming 169

Rock Shops 173

Campgrounds & Motels 190

Bibliography 237

Introduction

Rock and mineral collecting is a fast-expanding hobby, appealing to people of all ages and occupations. Within the last 20 years, tens of thousands of people have become attracted to this exciting activity and enthusiastic collectors are met today in practically every community.

Since this hobby has expanded so rapidly it certainly proves that it possesses some very strong points. Indeed, mineral and rock collecting does have exceptional advantages.

First of all, it is an outstanding family hobby, carried on primarily out-of-doors. Hiking or camping can be a part of this hobby. Family weekends, holidays, or vacation trips can be planned to include or be devoted to collecting.

It is a hobby that can be pursued to any degree a person desires. It can be enjoyed by a casual collector who just studies interesting rocks or it can be a stimulating scientific hobby for those interested in mineralogy.

It is an outdoor hobby that also has indoor opportunities, even for shut-ins. Pleasant evenings or days of bad weather can be spent examining specimens, labeling, cutting specimens, arranging displays, or mapping or planning future field trips.

Perhaps the best point of all is that being a collector of rocks and minerals contributes to meeting people and establishes new friendships with those who share in this popular hobby.

Rocks and Minerals

OF THE WESTERN UNITED STATES

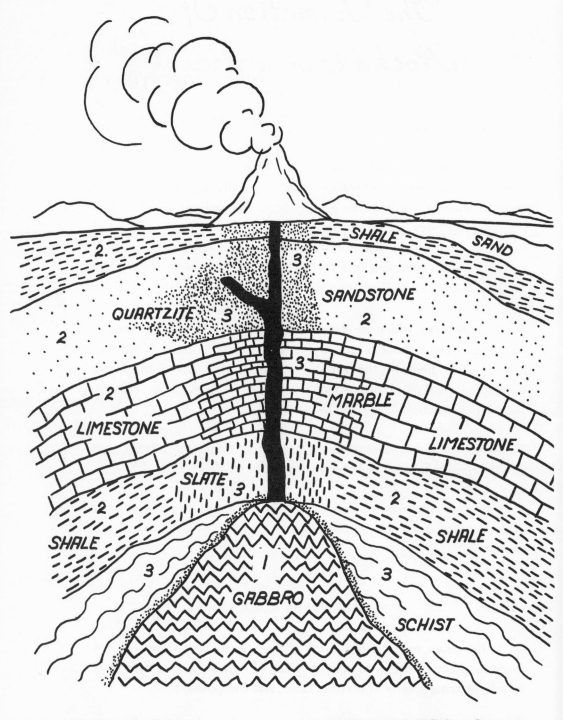

1. IGNEOUS ROCKS   2. SEDIMENTARY ROCKS
3. METAMORPHIC ROCKS

The Formation Of
Rocks & Minerals

Rocks are, essentially, the building materials of which our earth is constructed. Some are composed of glasses or organic materials, that do not have discrete materials. Some rocks consist of a single mineral while others contain several minerals or were originally formed from older rocks in which these minerals were present. All rocks are the result of geologic processes acting during the ages, building certain rocks and breaking down others. The normal rock cycle, which leads from molten rock to igneous rock and to sedimentary rock, followed by a metamorphic stage, gives rocks their infinite variety.

Igneous rocks, which are classified by their texture, mineral content, and origin, come from magmas, found deep in the earth. When this magma cools beneath the surface, intrusive rocks are formed and develope typical structures which later may be exposed by erosion. Magma that reaches the surface forms extrusive rocks.

Examples of intrusive rocks are pegmatite, syenite, peridotite, and gabbro. Granite is the best known of the deeper igneous rocks.

The most common member of the extrusive rock group is lava, but other familiar members include obsidian, rhyolite, pumice, basalt, and andesite.

SEDIMENTARY ROCKS

Sedimentary rocks, which have all been made of materials moved from the place of origin to another place of deposition, are extremely varied, differing greatly in texture, color, and composition.

Unconsolidated sand or mud is generally referred to as a sediment, while the consolidated materials are known as sedimentary rock. Those rocks which are made of grains or particles are called clastic. Other sedimentary rocks are of chemical or organic origin and most are formed in layers which often contain fossils.

Sandstone, shale, and limestone belong to the sedimentary family.

METAMORPHIC ROCKS

Metamorphic rocks are those which have been changed. The changes may bring about a new crystalline structure, may be barely visible in some rocks, while in others it is so great that it is impossible to determine what the original rock was. All rocks can be metamorphosed, which results from heat, pressure, or permeation by other substances. The heat and pressure increase with depth in the earth's crust and also has resulted from crustal movements or igneous activity.

Slate, marble, and quartzite are simple metamorphic rocks while phyllites and schists represent the more highly metamorphosed rocks.

NOTES OF INTEREST

Obsidian is formed when rhyolitic lava is quickly chilled.

Rhyolite lava blown to a spongelike consistency by the release of gases forms pumice.

Igneous rocks have long been known to be associated with metal ores.

Corals, crinoids, mollusks, and worms make up the minerals in limestone.

Marble is recrystallized limestone.

Placers are sedimentary deposits in which gold, gems, and other heavy durable minerals are concentrated.

One small handful of dune sand may contain several dozen different minerals.

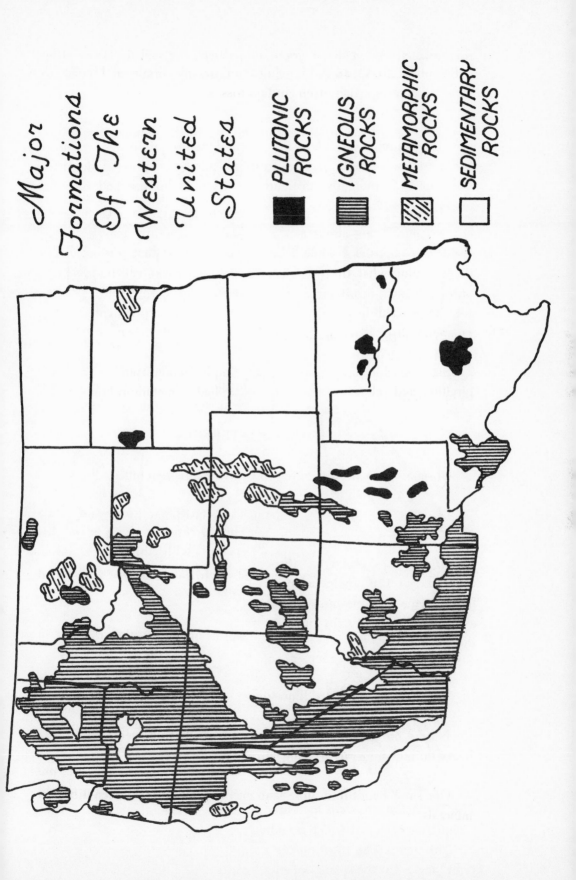

Major Formations Of The Western United States

PLUTONIC ROCKS

IGNEOUS ROCKS

METAMORPHIC ROCKS

SEDIMENTARY ROCKS

The Six Crystal Systems

The common crystal forms or combination of forms are shown below for each system. Crystal form is often used in mineral identification but it is the most difficult characteristic to use and one that requires considerable study.

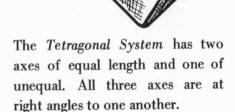

The *Isometric System* includes crystals which have three axes of equal length and are at right angles to one another.

The *Tetragonal System* has two axes of equal length and one of unequal. All three axes are at right angles to one another.

The *Orthorhombic System* has crystals with three axes, all at right angles, but all of different lengths.

The *Hexagonal System* has three equal axes at 120° angles arranged in one plane and one more axis at a different length at right angles to these.

The *Monoclinic System* has three unequal axes, two of which are not at right angles. The third makes a right angle to the plane of the other two.

The *Triclinic System* has three unequal axes, but none form a right angle with the others.

Minerals According To Hardness

Listed below are the more common gem stones and minerals according to the simple test of hardness.

2-3	3-4	4-5	5-6
Borax	Azurite	Colemanite	Apatite
Chalcocite	Bornite	Fluorite	Kyanite
Cinnabar	Calcite	Iron	Limonite
Copper	Chalcopyrite	Manganite	Obsidian
Gold	Crysocolla	Scheelite	Opal
Halite	Cuprite	Wolframite	Rhodonite
Silver	Howlite		Smithsonite
Sulphur	Malachite		Thomsonite
Tungstite	Rhodochrosite		
Wulfenite	Serpentine		

6-7	7-8	8-9
Amazonite	Aquamarine	Corundum
Amazonstone	Beryl	Ruby
Benitoite	Golden Beryl	Sapphire
Bloodstone	Spinel	
Chrysoprase	Topaz	
Epidote	Tourmaline	
Feldspar	Zircon	
Garnet		
Jade		
Jasper		
Marcasite		
Moonstone		
Petrified Wood		
Quartz		
Turquoise		

 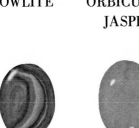   

| HOWLITE | ORBICULAR JASPER | RHODONITE | PETRIFIED WOOD | SNOWFLAKE OBSIDIAN |

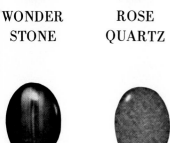    

| WONDER STONE | ROSE QUARTZ | BANDED AGATE | CALIFORNITE | RED JASPER |

  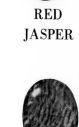 

| AGENTITE | BLUE QUARTZ | RHODONITE AUST. | LAVENDER JASPER | PALM |

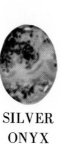  

| OLDSTONE | PEACOCK OBSIDIAN | SILVER ONYX | OOLITE | TURRITELLA |

  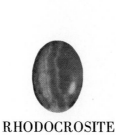  

| SPIDER WEB HOWLITE | AGATE | RHODOCROSITE | MOSS AGATE | GLORYSTONE |

APACHE TEARS

COLEMANITE

GEODE

JASPER

LEAD AND SILVER

PETRIFIED WOOD

QUARTZ CRYSTALS

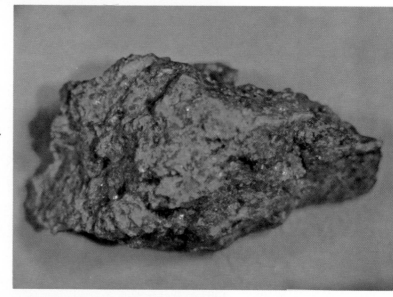

SULPHUR

TURQUOISE

Becoming Familiar With Rocks And Minerals

The beginner has much to learn and his knowledge is best gained through observation and study.

Visiting public museums that feature exhibits of rocks and minerals is one of the best ways to become familiar with the field.

A standard textbook on geology and an introductory book on mineralogy would prove helpful to the beginner. Public libraries offer such books in the field of geology and mineralogy, ranging from beginners manuals to more advanced studies.

Subscribing to monthly mineral magazines would provide current sources of information.

Joining a local gem society is an ideal way to become familiar with rocks and minerals, in addition to meeting other collectors and learning from those more experienced. Remember, the experienced collector is generally only too happy to share his knowledge.

From an early and continuing study of rocks and minerals and related texts, the collector can increase his knowledge of the physical properties that hold the key to successful identification. The physical properties of minerals may be readily observed by the eye and touch, or lie within the scope of a simple test.

Color, for example, is perhaps the most determinable physical property of a mineral, but at times it can also be misleading. The beginner will soon learn from experience to always check the color of a freshly broken surface. Time and the elements often change the exposed mineral surfaces.

Streak is another physical property used in mineral identification that is based primarily on color. Minerals often vary in their exterior coloring, but by drawing a part of the specimen over the surface of a piece of unglazed porcelain, the true color will be revealed.

Luster, which is also helpful in identifying minerals, refers to the natural appearance of a specimen's surface as it reflects when viewed under a bright light. Vitreous (glassy), adamantine (brilliant), resenous (waxlike), greasy (oily), and pearly (like mother of pearl), are terms used to describe luster.

Hardness is a characteristic often used in determining a mineral's identity. This is done by scratching the unknown mineral with a mineral of known hardness. With this method an unknown is soon bracketed by the mineral that scratches it and one that won't, thus approximating the relative hardness. A mineral on the Moh's Scale of Hardness shown below, will scratch anything equal to it or any mineral specimen of a lower number.

1. Talc	6. Feldspar
2. Gypsum	7. Quartz
3. Calcite	8. Topaz
4. Fluorite	9. Corundum
5. Apatite	10. Diamond

Hardness of many minerals often varies. This is commonly true when a variety of specimens from different localities are tested.

Specific gravity, which simply refers to the weight of a mineral in relation to the weight of the same volume of water, is an important physical aspect of minerals. The determination of this property is not easy for the amateur as the lack of purity of minerals will often lead to inaccuracies. Most collectors give an estimate of the weight by holding the specimen in his hand. Experience will give a surprising ability to estimate closely.

Cleavage, which has to do with the way a mineral splits along plane surfaces, is considered another important aid to identification. This property is the result of a precise pattern of atoms in regular layers, whose cohesion is weaker in certain directions than in others. Cleavage is not easily judged by the beginner as it is often difficult to tell the face of a crystal from that of a fresh, perfect cleavage surface.

Not all minerals will show cleavage, but experts say it can be best recognized by small steplike surfaces on the outside in preference to internal cracks.

Fracture, also used in identifying minerals, is when a specimen breaks in irregular directions, much like shattered glass. Most rocks and minerals show fractures and fresh fractures will show the specimen's true coloring. Such breaks, which appear differently in minerals, are generally defined as uneven, rough, hackly, splintery, earthy, or conchoidal. A conchoidal fracture, which is smooth, circular, and concave, is normally seen in obsidian. Copper and silver generally have a hackly fracture.

Magnetism occurs in some minerals, but magnetite and pyrrhotite are the most common. These minerals are often easily identified by the use of an alnico magnet or a common horseshoe.

Minerals possess a wide range of physical properties other than those already described. Final identification of difficult minerals often requires chemical tests, optical examination, or X-ray photography. These complicated procedures are clearly and fully explained in more advanced textbooks on mineralogy.

For further reference on rocks and minerals, the following books are especially recommended:

Berry, L.G. and Mason, B. *Mineralogy, Concepts, Descriptions, Determinations.* San Francisco: W.H. Freeman & Co., No Date.

Dana, James Dwight and Dana, E.S. *The System of Mineralogy.* New York: John Wiley & Sons, Inc., 1944.

English, George Letchworth and Jensen, D.E. *Getting Acquainted With Minerals.* New York: John Wiley & Sons, Inc., 1958.

Pough, Frederick Harvey. *A Field Guide to Rocks & Minerals.* Boston: Houghton Mifflin Co., 1955.

Descriptions Of
Rocks & Minerals

AGATE

Agate belongs to the chalcedony family, which means it is made of the silicas of both quartz and opal. Agates offer a great variety of colors and patterns, but both are often hidden by a dingy, rough coat. In moss agate, a black or dark brown mineral coloring is spread through a clear, light-colored quartz, resembling dark figures of moss, ferns, trees, or miniature landscapes.

AZURITE

Azurite, forming under conditions identical with those of malachite, is usually associated with malachite. It is commonly found crystalized in large well-formed deep blue crystals or in rosettelike aggregates. Azurite is considered rarer than malachite, but its stains on rocks generally serve as a guide to its location.

BERYL

Beryl, a silicate of beryllium and aluminum, is one of the major gemstones. When blue or a bluish green, it is aquamarine, when green, it is emerald, and when pink it is morganite. Common beryl, which often occurs in large rough crystals, is used industrially.

BORNITE

Bornite, occurring with the other copper sulphides, is an important ore of copper. Crystals are rare, but regular masses are readily recognized by their brownish bronze color on a freshly fractured surface. A purplish tarnish gives bornite the miners name of "peacock ore."

CHALCEDONY

Chalcedony may be fibrous or granular, but its surface tends to be botryoidal and often smooth and translucent. Agate is a banded chalcedony, carnelian is red chalcedony, and chrysoprase is green. The bloodstone is also green but with red spots that resemble drops of blood. The most commonly found colors of chalcedony are brown, green, grey, black, and white.

CHERT

Chert, a compact silica rock much like flint, is often found as a hard, red rock with white veins running through it. Some specimens may be white or light grey.

CINNABAR

Cinnabar, the chief ore of mercury, is associated with native mercury, realgar, opal, quartz, and barite. The soft, but heavy mineral is a bright red to brick red color and is generally found in shallow veins and rock impregnations.

COLEMANITE

Colemanite is commonly found in distinct crystals, usually well developed and more or less equidimensional, but it is also seen in granular masses. Colemanite is a borate mineral, formed in old lake beds in southern California.

CORUNDUM

Corundum resembles many silicate minerals, showing color bands and bronzy luster on its basal plans. Harder than any other natural mineral, except diamond, corundum is transparent to translucent, often triboluminescent. It is generally found in igneous rock, and is associated with spinel, garnet, and high-calcium feldspar.

CHRYSCOLLA

Chryscolla, a minor ore of copper, is often used as an ornamental stone and commonly substituted for turquoise. Chryscolla, like the turquoise, varies in color between a greenish and bluish hue, but it is not valuable. It usually occurs in the upper part of copper mines and when impure, often becomes brown or black. In this drab state it is generally known as "pitchy copper ore."

CUPRITE

Cuprite, often called ruby or red copper, is the only copper ore that is red. The rich ore often looks much like cinnabar, but it has a brownish red streak as compared to the bright red streak made by cinnabar. Cuprite, with its dark red metallic luster and common spots of green and blue, flecked with copper bits, is generally found near the surface of the earth.

FLUORITE

Fluorite, commonly found in cubes, is harder than calcite, but softer than quartz. A common vein mineral, fluorite often accompanies the ore minerals. Octahidral and complex crystals of this popular mineral are considered characteristic of high-temperature fluorite, while the cubic crystals are of a low temperature occurrence. Fluorite may be colorless, white, brown or black.

GARNET

Garnet is a name given to a family of minerals displaying different colors such as yellow, black, green, brown, and white. All have similar crystal habits, but vary in mineral make-up and other properties. Although easily confused with other gemstones, garnets do have a typical crystal with eight even sides. Garnet is often found in granite, but grains also appear in sands of stream beds and seashores.

JASPER

Jasper, belonging to the quartz family, is not pure quartz. The impurities from minerals that have joined the quartz, give jasper its rich colors of red, yellow, brown, green, and blue. These colors often combine in speckled, mottled, and banded varieties. Ribbon jasper has parallel stripes of different colors that often have bands of rock crystal inside.

KAOLINITE

Kaolinite, the best known of the clay minerals, is formed by the weathering of that containing large proportions of feldspar. The thin, white crystal plates of this clay mineral are rarely large enough to be seen by the naked eye, but masses of kaolinite are easily recognized by a peculiar earthy odor.

LIMONITE

Limonite is actually the coloring matter of soils, formed from iron minerals at surface temperatures. Its fine crystallized specimens or large cubes are colored brown black to ochre yellow.

MAGNETITE

Magnetite, an important ore of iron, often forms small grains in igneous rocks, which after weathering, are often concentrated into black beach sands. The gray or black magnetite is also found well crystallized in pegmatitis and high temperature veins.

MALACHITE

Malachite is the commonest staple of the secondary ores of copper, generally forming near the ground surface. The light to dark green colored ore usually forms fibrous crystals and masses, while single crystals are quite rare.

MICA

Mica has the remarkable property of peeling into thin sheets, which are easily bent, and when released, swing back to their original position. There are several kinds of mica, all having different colors, because of the variations in their chemical formulas. Lepidolite mica is a pink or purple color, muscovite mica is white or green, biolite mica is black or dark brown and phlogopite mica is a yellowish brown.

OBSIDIAN

Obsidian is a glassy rock which has not yet crystallized, because it cooled too quickly for any atoms to group into arrangements of the minerals. Obsidian is generally found where there has been volcanic activity in recent times.

OPAL

An opal, characterized by a play of rainbow colors from what is essentially clear material, is a precious stone, while the clear orangered opal is a fire opal. Common opal, which has no real value, is often highly fluorescent. A clear colorless opal is as ashyalite. Opal is generally found in recent volcanics, deposits from hot springs and also in sediments.

PYRITE

Pyrite, frequently associated with all sorts of metal ores, is often mistaken for gold and therefore is popularly known as "fools gold." The tarnished sulphide is also commonly confused with chalcopyrite, but its hardness is distinctive. The light, yellow colored pyrite is harder than gold, but very brittle.

QUARTZ

Quartz, considered to be the most common mineral, can be found almost anywhere. Most commonly found transparent and colorless, the crystals of this hard mineral are easily recognized by the hexagonal pattern or their typical points. The high temperature veins are usually coarsely crystallized, while low temperature veins in sedimentary rocks are finer grained.

RHODOCHROSITE

Rhodochrosite, like rhodonite, is a beautiful pink color that distinguishes it from all other manganese ores. Its crystals usually have six faces, slightly curved and somewhat triangular. This mineral is not considered common, but may be found in veins of gold, lead, silver, and copper ores.

RHODONITE

Rhodonite, a fine, rose-pink manganese ore, is not a common mineral, but may be found as pebbles or rocks in stream and beach sands. The manganese gives rhodonite its light pink or sometimes brown coloring, but impurities may give it a yellowish or greenish color.

SERPENTINE

Serpentine, a common mineral, is usually relatively soft and a dark greenish color with snakelike patterns of variegated patches of blue darker and lighter colors. Serpentine may be either platy or fibrous, but often it resembles green marble. Crystals of this mineral are unknown.

TOPAZ

Topaz is commonly found crystallized, often in large free growing transparent crystals. Although usually thought of as a yellow stone, topaz may occur colorless, blue, yellow-brown, or a pinkish brown.

TOURMALINE

Tourmaline has the greatest variety of colors and tints of all the gemstones. Its large crystals are often black, blue, pink, red, green, and yellow. Occasionally different colors form in a single crystal. Tourmaline is generally found in granite, limestone, or related rocks and in cavities of pegmatites.

TURQUOISE

Turquoise, a mineral found almost invariably in arid climates, is often confused with an iron-stained fossil bone known as odontolite. Commonly found where rocks have been deeply altered, turquoise is usually a light blue to a greenish blue color. When the matrix, or adjoining rock, is present as a delicate veining, it creates a spider weblike pattern, desired by many as a sign of genuineness.

WULFENITE

Wulfenite is, to a small extent, an ore of molybdenum, but its crystals have been considered the loveliest in the mineral kingdom. Colors run from brilliant red to orange, yellow, and brown, but some may be gray or brown. Wulfenite, generally resulting from the decomposition of other minerals, is associated with vanadinite and pyromorphite.

NOTES OF INTEREST

Many topaz crystals will fade when exposed to much sunlight.

Topaz colors are often altered by heat to produce more salable hues.

The green patina often seen on copper and bronze is actually a thin tarnish of malachite.

Pyrite is of little value for its ore content, but it is often rich in gold and copper as well as being a source of sulfur.

Feldspar and quartz occur together in a rock called pegmatite.

Calcite comes in more than 300 different crystal forms.

Tools and Equipment

To collect rocks and minerals efficiently, a few specialized tools are essential. Too many collectors go into the field with inadequate equipment, and failure to free a prize specimen often results.

The most essential tool in the field is the prospector's hammer, which is available in different weights and with handles of wood or metal. Wooden handles absorb more shock from hammering on rock than will the metal ones.

A small sledge with about a 5-pound head is useful for breaking up large rocks and also for driving chisels into rocks.

A broad-faced shovel, with a short handle is often required for digging in dumps or for opening the ground.

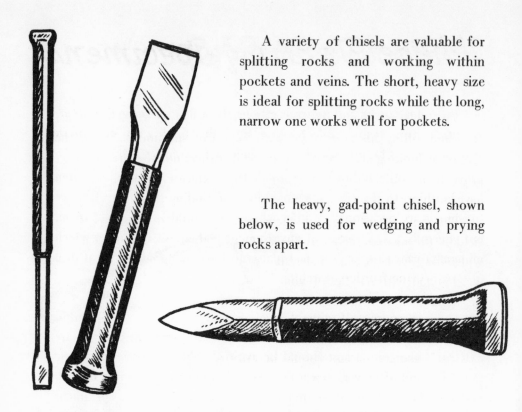

A variety of chisels are valuable for splitting rocks and working within pockets and veins. The short, heavy size is ideal for splitting rocks while the long, narrow one works well for pockets.

The heavy, gad-point chisel, shown below, is used for wedging and prying rocks apart.

The gem scoop, shown below, is for retrieving hard-to-reach rocks that would otherwise be impossible to grasp. This same tool is also used for raking rocks from water.

Other items of importance for collecting in the field are: a pocket knife or nail for testing hardness, a pocket magnifier for examining specimens, and a compass. The compass is used as an orientation and direction finder for hikes over unknown country. It may also be used to test for magnetism. By holding suspected minerals near the compass, needle deflection will indicate a degree of magnetism.

A knapsack or gunny sack will also be required for carrying specimens and plenty of old newspapers for wrapping.

Where To Find Specimens

The odds are against the novice merely selecting a likely looking area and filling his gunnysack with beautiful specimens. Often the promising-looking areas prove barren while others not so attractive may prove to be outstanding. This proves that exterior appearances often have nothing to do with the possibility of finding specimens. Choice specimens are occasionally found where you would least expect them, but generally a knowledge of environment and conditions under which minerals form, as well as logical places to look, eliminates a great deal of territory and fruitless searching.

The best sources of mineral specimens are mines and their dumps. The old abandoned mines, with their rotten timbers and foul air, are extremely dangerous and should be avoided. The mine dumps are the safest bet and often very rewarding. These dumps are composed of rock materials brought from far underground, therefore making specimens available to the collector that he could not otherwise hope to reach. Specimens found on the surface of a dump may be worthless or badly damaged, but a little digging may be very rewarding. Remember, these dumps are composed of different layers piled one atop the other and although the surface may appear fruitless, prize specimens may be found by the collector who is willing to exercise a shovel.

Rock quarries, gravel and clay pits, highway, railroad, dam cuts, tunnel and foundation excavations, temporary lakes in arid regions, beach sands and stream beds are all promising places for a collector to search.

Often it is possible to obtain leads to mineral localities by asking other collectors or corresponding with those belonging to societies in other areas or states.

Visiting rock shops and talking with the owners and customers often result in gaining valuable information.

Another source of leads for mineral localities is often found on the labels accompanying specimens in private collection or public displays.

Safety in the Field

Rock and mineral collecting, like many outdoor activities, requires an understanding of terrain and of possible problems that may be encountered. Poisonous snakes, for instance, are native to most areas of the western United States and they can be a real threat to the careless collector. Never step or place your hands into unseen places. Caves, old mines, buildings, abandoned autos, wood piles, rocks, and brush are always possible snake havens (black widow spiders and the poisonous scorpions also favor such places). In areas suspected of concealing snakes, make some noise to advertise the fact that you are coming, or better yet, avoid such places. Keep in mind, that no specimen is worth endangering your health or life.

When collecting around old abandoned mines or buildings watch where you are walking. Mine shafts at the ground level are often overgrown with brush. Never walk on innocent-appearing old boards as they may cover an old well or mine shaft. These types of places should always be avoided at night.

If you go into a remote collecting area, always let someone know where you are going and give them an approximate time as to when you expect to return. Travel with a companion when possible. Be conscious of landmarks to avoid becoming lost. Keep in mind that the beautiful wide-open spaces can be hostile to man. Before entering a desert, listen to weather conditions on local stations to avoid being caught out in a rain or sandstorm. Avoid entering a desert during midday, especially during the summer months. Never drink from streams or pools, especially those around mines. Mountain streams are generally safe to drink from but it is best to carry a canteen of water.

Proper clothing is also very important. During hot weather always wear a hat to protect your head from the sun's burning rays. Leather boots, with rubber soles for good footing, will protect the lower legs and ankles from sharp rocks and brush. Boots should always be worn in areas suspected of having poisonous snakes.

Specimen Sizes to Collect

The selection of a specimen size to collect is purely a personal matter, but most collectors quickly find practical advantages in focusing their collection within the bounds of an approximate size. There are several reasons for this. One is that the neat format of a collection in which the specimens are all of an approximate size is certainly more attractive than a hodgepodge of sizes. Another factor is the cost of mineral specimens. Large specimens naturally cost more than small ones and therefore the average pocketbook can better afford the smaller specimens. Another point that should be mentioned here is that you don't have to have large specimens to have quality. The finest specimens are often only 1½ x 1½ x 2 inches. These miniatures also offer ease of storage, display and availability of many varieties at modest prices. It is easy to see why this size is considered the most practical for home collections.

Protecting Specimens

Collected specimens, especially crystals, should be properly protected or wrapped as they are collected in the field. The little extra time it takes to wrap specimens will prevent ruined or damaged pieces. Old newspapers are agreed to be the most inexpensive and satisfactory wrapping medium for most specimens, but small paper bags, cleansing tissue, and tissue paper may also be used.

Wrap specimens by folding the paper as many times as possible for protective cushioning. The extra-special or delicate specimens should be carefully wrapped in tissue paper or cotton and then placed in the knapsack pockets.

When packing specimens in the knapsack or gunnysack, remember to always place large, heavy pieces in the bottom and the smaller or delicate pieces on top.

Field Notes

Most serious collectors have the habit of making field notes for collected specimens and enclosing them with the pieces as they are wrapped. Important data should not be left to memory.

Use a small pad or notebook to write down such important information as location, date, name, or description of the specimen and any related remarks applicable to observations of the locality in general.

Cleaning Specimens

The frequent washing of many collected specimens should be avoided. Scouring agents can ruin soft specimens. Very hot or cold water should also be avoided as some specimens, plunged into such baths, will fracture or even shatter. Don't forget, too, that many minerals are soluble in water.

Stains in mineral specimens are often removed by chemical action, but the correct method to use for the particular mineral or combination of minerals involved is often critical.

Each mineral has an individuality of its own and the technique of cleaning often requires considerable time to learn. Until such knowledge has been acquired it may be safest to consult veteran collectors or a professional mineral dealer. Don't risk a ruined specimen.

Housing and Display

Just as important as collecting rocks and minerals is housing or displaying them in an attractive, practical manner.

Prefabricated specimen boxes, available with 100, 75, 50, 24, and 18 individual compartments, are useful for housing small specimens. These lightweight but sturdy cardboard boxes may be purchased from scientific supply houses or mineral dealers.

Glass cases are preferred by most collectors for displaying since the minerals show up so well and they are protected from dust. New cases can be purchased from showcase companies, but more often secondhand cases are used. Although used cases often require some refinishing, or glass replacement, the savings are well worth considering.

How you display your collection is a matter of personal preference, but to show specimens at their best, there are a few important factors to remember. Never overcrowd a collection. Arrange specimens neatly and try to keep a good balance in the display. Place large pieces on the bottom or to the back of a display. Specimens that are difficult to stand up in a position to show them off to the best advantage should have bases made from styrofoam, wire, or wood. Speciman bases add a professional touch to a collection, but there is a danger of overuse where base material overwhelms the specimens. The display should be well illuminated by incandescent bulbs as fluorescent tubes often disguise a mineral's true coloring. Last but certainly not least, keep the displayed specimens dusted. A large, soft bristle paintbrush is perhaps the handiest tool for dusting specimens.

Gem And Mineral Locations

Collecting Localities

Many of the gem localities described on the following pages are already publicized and well known to the veteran collector. Some have been thoroughly worked over, but new finds are constantly being made in the same general areas. For every ounce of quality gemstones there are tons of common rock, therefore the successful collector must often search a little farther or dig a little deeper.

Many states have laws prohibiting or limiting the collection of petrified wood and the federal government has enacted laws regulating the collecting of such specimens on public lands. These regulations are in the formation stages and therefore are subject to changes. The many popular rock and mineral magazines are the best source for current information.

Rock or mineral specimens should never be collected in national parks, national monuments or any areas that are administered by the National Park Service. It is against federal regulations to collect in these areas.

Private claims and most restricted areas are generally posted, but many owners will grant the collector permission to gather specimens. Some may charge small fees and limit the amount of material collected. Always obtain permission to hunt on private property and respect the rights of the owners. The fact that localities are listed in this book does not mean that the reader can trespass without the proper authority.

Explanation Of Map Symbols

paved highways
towns & cities
improved & unimproved roads
mineral & mine locations
approximate mileage between towns & dots.

43

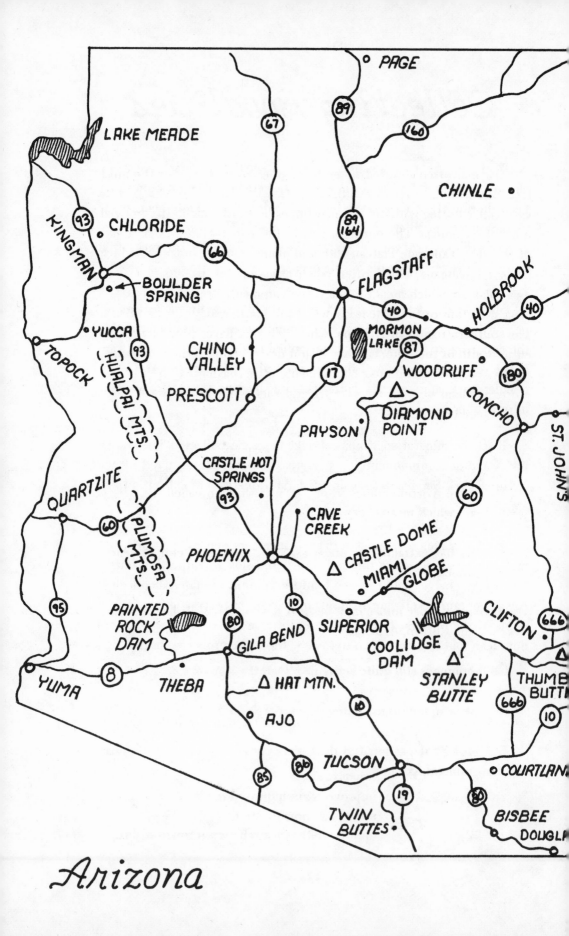

Arizona

Arizona

Arizona, often referred to as the Copper State, is one of the most heavily mineralized regions in the entire world. Almost every part of its vast land is dotted with old and new mines, most of which afford prized mineral specimens. The northern part of this mineral rich state is a part of the great Colorado Plateau with arid plains, colorful buttes, and high timbered plateaus which often reach elevations of 10,000 feet. This great plateau, which breaks into a vast escarpment, stretching across the central section of the state is known as the Tonto and Mongollon Rims. The southern region of Arizona, where the Gila River flows across the entire width of the state, is a land of hot desert plains, broken by ranges of dry, eroded mountains. The immense state, having 113,575 square miles, is a land of contrast with its arid cactus-filled deserts and lofty snow-capped mountains.

NOTES OF INTEREST

The name Arizona comes from the Spanish, who called the region "arid zona," which means "dry belt."

Arizona has extracted as much as 429,200 tons of copper yearly from her mines.

Open-pit copper mines may be seen today at Ajo, Bisbee, Jerome, Morenci, and Clifton, Arizona.

Gold mining is still quite active near Oatman, Arizona.

Cuprite is an important source of copper at the Bisbee mines.

The geology department at the University of Arizona in Tucson has a notable mineral exhibit.

Apache County — Concho area, E toward St. John's, moss agate, Milk Ranch, NW of Concho, petrified wood, St. John's area, N of town, banded agate and petrified wood, Lyman Lake, S of St. John's petrified wood, Woodruff area, petrified wood, Joseph City, area NE of town, petrified wood.

⊕ Specimen Locations

Mineral Occurrences In Arizona

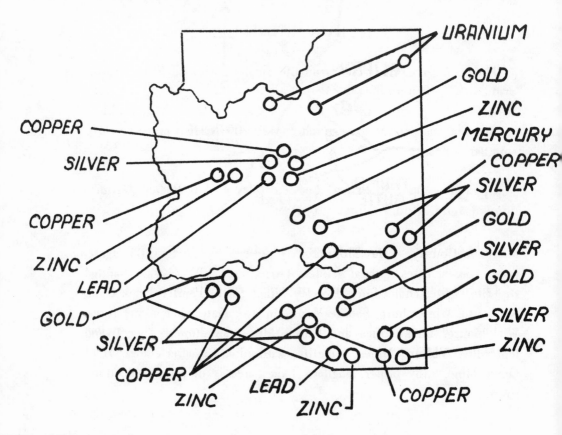

46

Cochise County — Bisbee area copper mines, bornite, azurite, malachite, native copper, Guadalupe Canyon, NE of Douglas, agate, opal, jasper, Skeleton Canyon, SE of Apache, Moss agate, jasper, opal.

Gila County — Globe district, Apache Mine, Vanadinite with descloizite, Gila River, S bank below Coolidge Dam, agate, jasper.

Graham County — Stanley Butte, S of Coolidge Dam, green and yellow garnet, Klondyke, Table Mountain copper mine, red vanadinite and yellow wulfenite, Ashurst, Box Canyon, quartz crystals, chalcedony roses.

Greenlee County — Clifton area, N and NE of town, chalcedony roses, agate, jasper, crystal geodes, Thumb Butte, jasper, chalcedony nodules, Apache Creek, fire agate, geodes, carnelian.

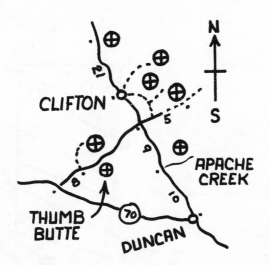

Maricopa County — Hat Mountain, N of Ajo, E side hwy 85, agate, jasper, desert roses, crystal geodes, Theba, area NW of town, banded rhyolite, agate, chalcedony, Rowley Mine, SW of Painted Rock Dam, wulfenite, Wintersburg, Saddle Mountain, fire agate, chalcedony roses, Wickenburg, Lucky Mica and Garcia Mines, spodumene, beryl, and lepidolite, Vulture Mines, minetite, wulfenite, Cavecreek area, E of Onyx Mine, ruby jasper, Saguaro Lake, area N of lake, chalcedony roses.

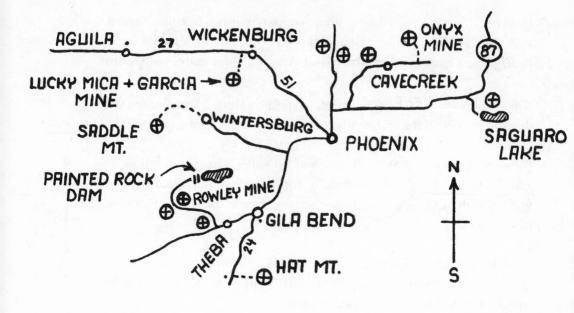

Mohave County — Chloride area mines, desclozite, endlichite, vanadinite, Boulder Spring, mine, amethyst, Yucca, area SW along W side of hwy 66, Goldroad, Sitgreaves Pass, fire agate, Bullhead City, area SW of town, agate, rhyolite, carnelian, Topock, area N of town, along both sides of hwy, jasper, chalcedony.

Pima County — Ajo area mines, azurite, chrysocolla, malachite, Casa Grande, Silver Reef Mine, amethyst, Wild Horse Pass, agate, Marana, Silver Hill Mine, azurite, malachite, rosasite with rhombs of smithsonite, Twin Buttes, Copper Queen Mine, malachite, bornite.

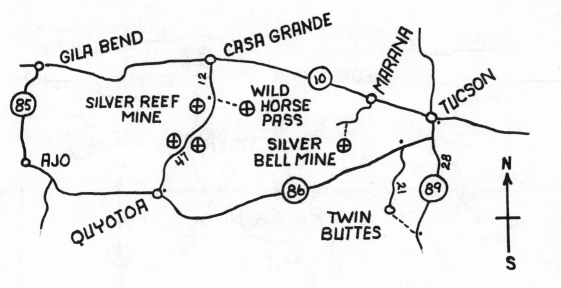

Pinal County — Superior, area W and SW of town, apache tears, Florence, area NE of town, amethyst, Mammoth, area N of town, E of hwy, chalcedony.

Santa Cruz County — Patagonia, area SE of town, chalcedony, jasper, geodes, ruby area NE of town, agate.

Yavapai County — Castle Hot Springs area, black tourmaline, quartz crystals, agate, chalcedony, Wickenburg, mines NE of town, quartz, amethyst, smoky quartz, Perkinsville area, agate, pink chert, Ash Fork, area NE of Cathedral Cave, agate, Bagdad, area mines, native copper, chalcocity, malachite, Burro Creek, NW of Bagdad, opal, chert, jasper.

Yuma County — Castle Dome area mines, fluorite, calcite, wulfenite, barite, quartsite, area S to Castle Dome, E side of hwy, quartz, crystals, crystal geodes, chalcedony, rhyolite, Cibola, river pebble areas to the north, agate, jasper, chalcedony, petrified wood, Martinez Lake, Red Cloud Mine, wulfenite, Wellton, Bentonite Mine area in Muggins Mountains, agate.

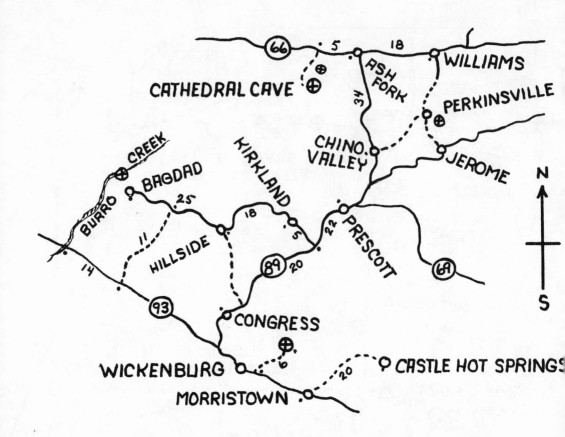

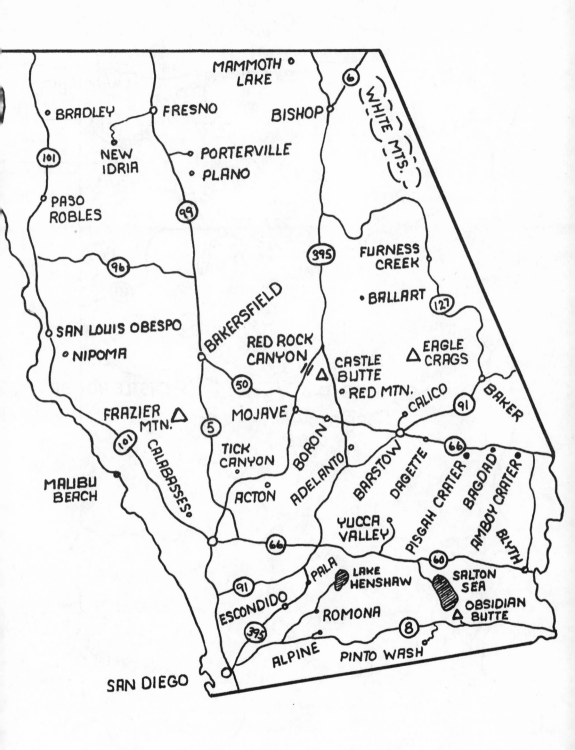

Southern California

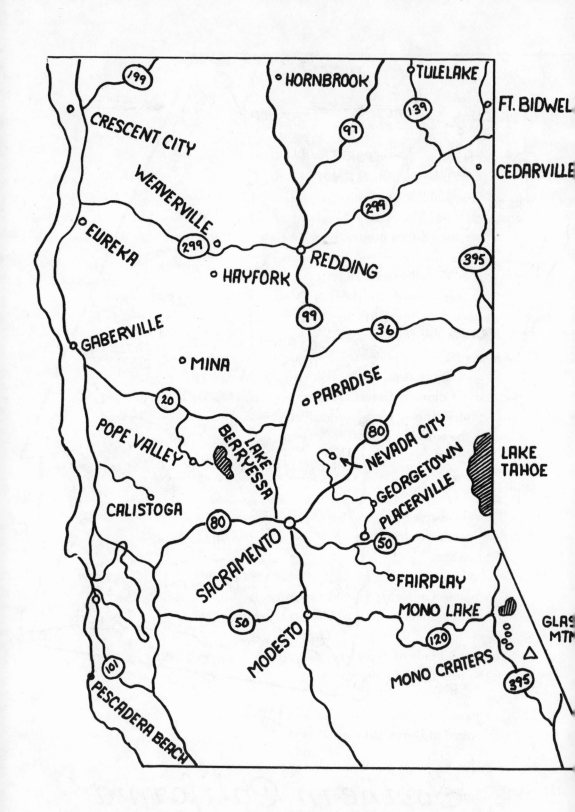

Northern California

California

California has more diverse topography and climate than any other state. Mt. Whitney, which towers to 14,495 feet, is the nation's highest point, while only a few miles away Death Valley sinks to 280 feet below sea level. The state's backbone is the towering Sierra Nevada Range near the eastern border, which extends southward for 400 miles from the Cascade Range and then forms most of Central Valley's east wall. The valley, a 450-mile alluvial trough, is walled on the west by the coastal ranges, closed to the north by the Klamath and Cascade Mountains and to the south by the Tehachapi Mountains. East of the Sierra Nevada lies the Great Basin region which is represented in the north by a high lava plateau. To the south there are barren mountain ranges and desert basins. In the southern section of the state lie the vast Mojave and Colorado Deserts. California, long famous for its Mother Lode Country mines, ranks among the leading gemstone and mineral regions of the world.

NOTES OF INTEREST

The coastal area of central California had many deposits of cinnabar and the state's contribution to the world's mercury production was quite impressive.

The name California is said to come from an imaginary island mentioned in a Spanish novel written in 1510.

About 60 kinds of commercially produced minerals put California among the first three mining states.

Colemanite was first noticed in Death Valley in 1883, while borax was discovered at Borax Lake in 1856.

Alpine County — Markleeville, area SW of town, banded lazuite, quartz, Silver City mine site, SW of Markleeville, petrified wood, marcasite.

Amadore County— Silver Lake, Bear Valley, SW of lake, petrified wood, Volcano, area NE of town, quartz crystals, Ione, Mooney Claim, SE of town, chrysoprase, Fiddletown area mining claims, rhodonite.

Butte County — Sterling City, Sawmill Peak, S of town, quartz crystals. Belden. area SW of town, californite; Oroville, Oroville Dam, area NW of dam, quartz crystals; Forbestown, mining areas, N. of town, rhodonite, rose quartz.

Coslio Chrome, San Luis Obispo, California.

Calaveras County — Valley Springs, area NE and E of town, moss agate. Murphys, area W of town, marble, area NE of Murphys, quartz, Copperopolis, area E of town, mariposite.

Del Norte County — Crescent City. beach areas N. of city. agate, jade pebbles.

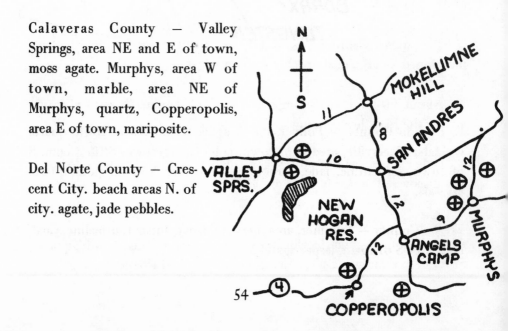

54

Mineral Occurrences In California

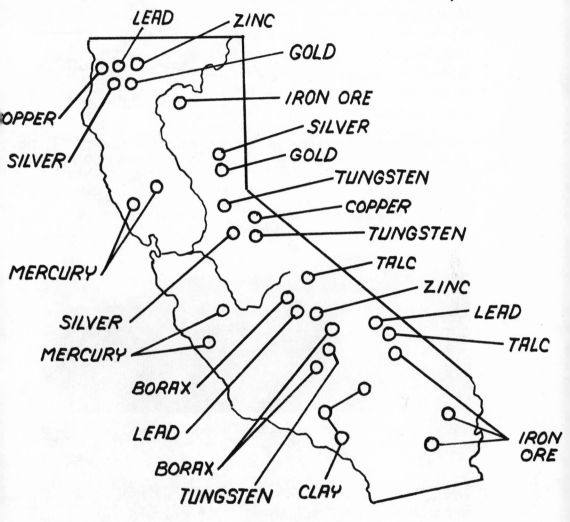

El Dorado County — Pollock Pines, areas W and SW of town, quartz crystals, Placerville, area N of town, jade, Georgetown, Stifle Claim, S of town, serpentine, jade, garnet, Volcanoville, N of Chiquita, quartz crystals.

Fresno County — Trimmer, area, beryl-feldspar, topaz-tourmaline, Coalinga, area S of town, jasper-agate.

Humboldt County — Trinidad area beacher, californite, jasper, carnelian.

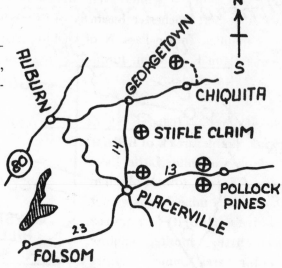

A striking feature of the California Coast Ranges is the steep up-ending of their strata as a result of strong folding and faulting at various times during the Cenozoic period. At Point Pedro, the up-ended strata are shales alternating with thin beds of sandstone of the late Cretaceous Age. *Courtesy C. W. Jennings, California Division of Mines and Geology.*

Imperial County — Winterhaven, area W of town, along both sides of hwy 8, jasper; Rochester Basin, N of Winterhaven, chalcedony roses, agate nodules. Indian Pass, N of Ogibby, agate, Jasper; Midway Well, mining claim E of town, turquoise; area S of Midway Well, E side of hwy, crystal geodes.

Inyo County — Ballarat, NE of Trona, marble; area E of Ballarat, epidote, wulfenite; Little Lake, area NE of town, obsidian; Darwin area mine dumps, garnet, pyrite, fluorite, bornite, malachite, apptite, epidote; Keeler area mines, azurite, malachite, chrysocolla, smithsonite, limonite; Lone Pine, mines NE of town, beryl, jasper, opalite, turquoise, wulfenite; Independence, Crystal Ridge, NE of town, quartz crystals; Deep Springs, Crystal Hill, SW of town, quartz crystals, Sulphur mine E of Deep Springs, variscite.

Kern County — Isabella Reservoir, N of Bodfish, mine NW of reservoir, quartz, garnet, scheelite; Woody, claim N of town, tourmaline, feldspar; Erskin Creek, SW of Bodfish, green garnet crystals; Tehachapi, claim NW of town, cinnabar; Rosamond, area NW of town, opal, jasper wood; Castle Butte, area NE of Mojave, jasper, agate, bloodstone; Boron, Boron Open Pit,

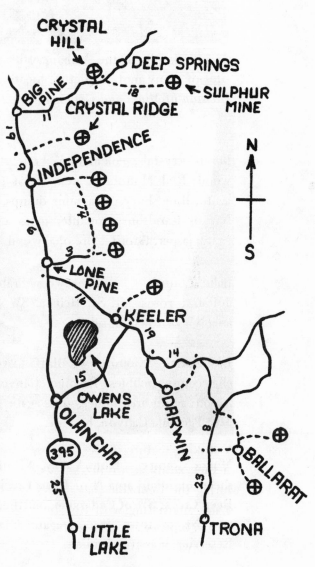

Granitic knobs of the Alabama Hills near the scenic road that traverses sites of many motion picture locations. *Courtesy M. R. Hill, California Division of Mines and Geology.*

borate crystals, colemanite; Lonely Buttes, SE of Mojave, petrified wood; Red Mountain, area NE of town, jasper-agate, saronyx, mose agate; Randsbury, area mine dumps, jasper, rhodonite; Sheep Springs, NW of Randsbury, opalite, moss agate; areas SW of Sheep Springs, agate, jasper, fluorite, fire opal wood.

Lake County — Middletown, Mirabel Mines, S of town, cinnabar, dolomite roses; Mt. St. Helena, SW of Middletown, red jasper; Cobb, area NW of town, obsidian.

Los Angeles County — Bluff Cove area beaches, agaten calcite, chalcedony pebbles; Pacoima Canyon, N of San Fernando, apatite, zircon; acton area mines, moss agate, geodes, bloodstone; Tick Canyon, N of Pacoima Canyon, howlite.

Modoc County — Plum Valley, NE of Davis Creek, area mines, golden sheen obsidian; area N of Davis Creek, in fill at road cut, jasper-agate; Deep Creek, SW of Cedarville, petrified wood, calcite crystals; Lookout, area NE of town, jasper, agate; Glass Mountain, SW of Clear Lake Reservoir, massive obsidian.

Large pits such as this one at Cherokee are common in many areas of
California, especially in the Sierra Nevada foothills, which were mined
for gold in the 1800s. These large pits were the result of hydraulic
mining, which is now prohibited. *Courtesy A. L. Sherriffs, California
Division of Mines and Geology.*

Mono County — Agnew Meadows, NW of Mammoth Lakes, hematite,
rhodochrosite; area NE of Mammoth Lakes, chalcedony, agate, obsidian
nodules. Parker Lake, W of Crestview, quartz crystals; Bodie, area NE
of town, opal, cinnabar, Crystal Hill, N of Bridgeport, quartz crystals.

Monterey County — Jade Cove, S of Lucia, jade.

Nevada County — Nevada City, mine NE of city, opal wood; Washington, local mine, green garnet; Meadow Lake, NE of Washington,
chalcopyrite.

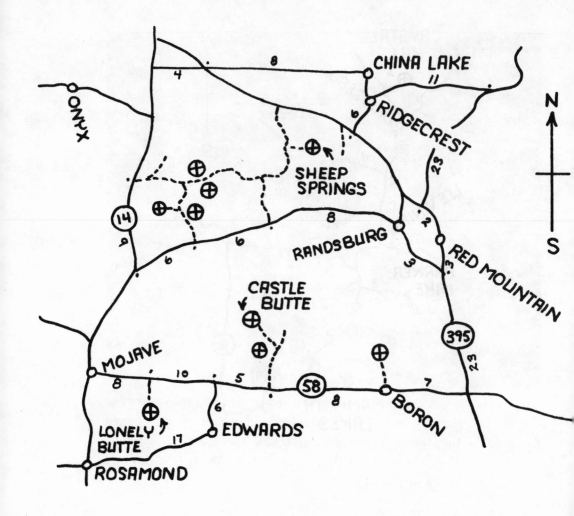

Pluma County — Genesee, area W and SW of town, quartz crystals, chlorite; Frenchman Reservoir, area SW of reservoir, rose quartz.

Riverside County — Coahuilla Mountain, N of Coahuilla, beryl, rose quartz, tourmaline; Bautiste Canyon, NE of Coahuila, smoky quartz; Lakeview, Mountain View Ranch, S of town, beryl, quartz crystals, tourmaline, garnet; Storm Jade Mountain mines, NE of Cottonwood Spring, agate, jade.

San Bernardino County — Owl Lake, S of Death Valley Nat. Mon., W of Salt Springs, agate; Baker, Toltec Mine, NE of town, torquoise; Afton, area S and SW of town, chalcedony, geodes, moss agate, jasper,

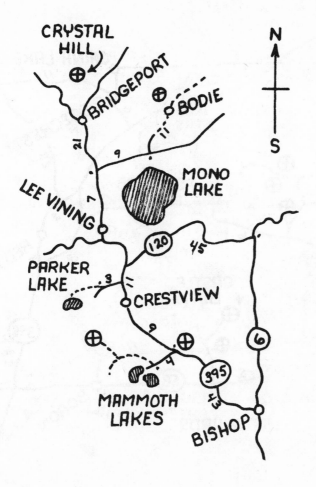

opalite, calcite clusters; Alvord Mountains, NW of Afton, chalcedony roses, jasper, agate, petrified wood; Mannix, area N of town, jasper, chalcedony, agate, calcite; Calico Mountains, NW of Calico, satin spar, jasper, agate; Owl Canyon, N of Barstow, jasper; Black Canyon, N of Hinkley, agate nodules; Newberry, area S of town, calcite geodes, agate nodules; Pisgah Crater, areas W and E of crater, agate, jasper; Ludlow Cady Mountains, NW of town, moss chalcedony, agate, jasper, quartz crystals, calcite, chalcedony; Ash Hill, SE of Ludlow, chalcedony roses; area NE of Ludlow, chalcedony roses, jasper; Bagdad, area N of town, chalcedony roses, apache tears; Marble Mountain, N of Cadiz, agate, topaz, garnet, marble, pyrites, hematite; Carsons Wells, SW of Needles, quartz crystals, chalcedony, agate; Chemehuevi Wash, SE of Needles, geodes; Whipple Mountains, NW of Parker, jasper.

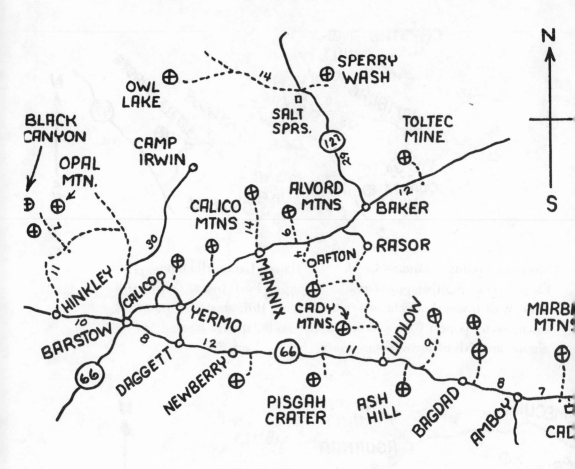

San Diego County — Chihuahua Valley, E of Oak Grove, tourmaline, lepidolite; Pala, Pala Gem Mines, tourmaline, garnet; Rincon, Victor-Mack Mines, garnet, beryl, tourmaline; Aguanga Mountains, N of Lake Henshaw stopaz, morganit; Mesa Grande, area mines, garnet, beryl, quartz crystals, tourmaline; Julian, San Felipe Garnet Mine, calcite, garnet, diopside; Romona, Surprise and Little Three Mines, topaz, smoky quartz, beryl, garnet; Boulevard area N and NE of town, garnet, beryl, mica, tourmaline; Pinto Wash, E of Jacumba, petrified wood, wonderstone.

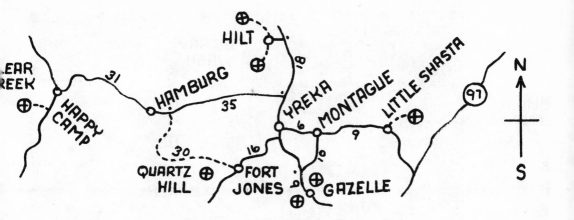

Siskiyou County — Indian Creek, N of Happy Camp, rhodonite, jade; Clear Creek, S of Happy Camp, rhodonite; Fort Jones, Quartz Hill Mine, W of town, rhodochrosite, chalcopyrite; Hilt, areas NW and SW of town, agate, jasper; Willow Creek, S of Gazelle, quartz minerals; Little Shasta, area NE of town, jasper, agate.

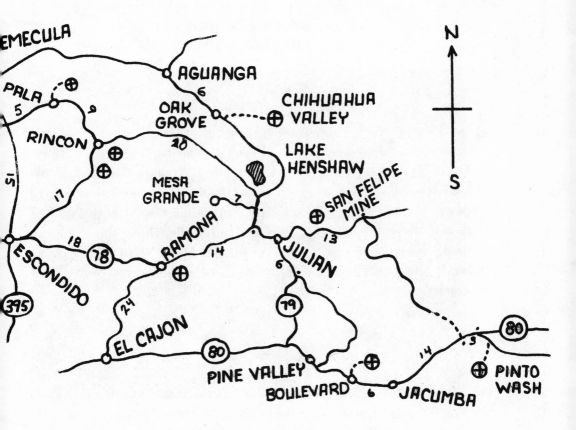

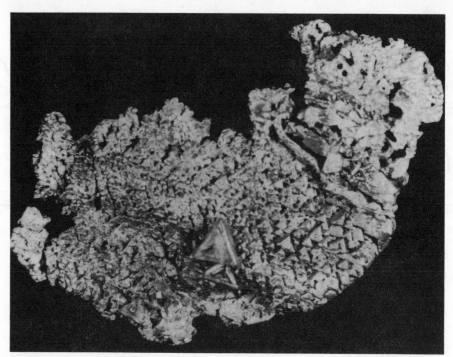

Famous Golden Bear nugget, on display at the California Division of Mines and Geology museum in San Francisco Gold is California's state mineral. *Courtesy M. R. Hill, California Division of Mines and Geology.*

Fossil elephant tusk from Santa Cruz County This specimen is on display in the California Division of Mines and Geology Museum in San Francisco. *Courtesy M. R. Hill, California Division of Mines and Geology.*

Trinity County — Hayfork, S of Weaverville, local mine, garnet; Mina, E of Garberville, area NE of town, jade, crocidolite.

Tulare County — ¾ Potterville, Deep Creek, SE of town, chrysoprase; Three Rivers, claim area N of town, garnet, epidote, indocrase, quartz crystals; Mineral King Mine, E of Three Rivers, epidote, tourmaline, limonite.

Ventura County — Camarillo, Count Park, area N of the park, opal, chalcedony roses, marcasite, agate; Faria Beach area, W of Ventura, jasper.

Serpentine is California's state rock. This specimen contains veins of chrystotile asbestos. *Courtesy J. T. Alfors, California Division of Mines and Geology.*

NOTES OF INTEREST

The summer months are best for collecting in most areas in the northern part of California, while late fall, winter, and spring are considered the best in most southern areas.

The California Academy Of Sciences in San Francisco, The Los Angeles County Museum, The San Diego Natural History Museum, and the Santa Barbara Museum of Natural History are but a few places in California to see notable mineral exhibits.

Some mines in the famed Mother Lode country of California are still producing.

Old Independence Stopes, Cripple Creek, Colorado.

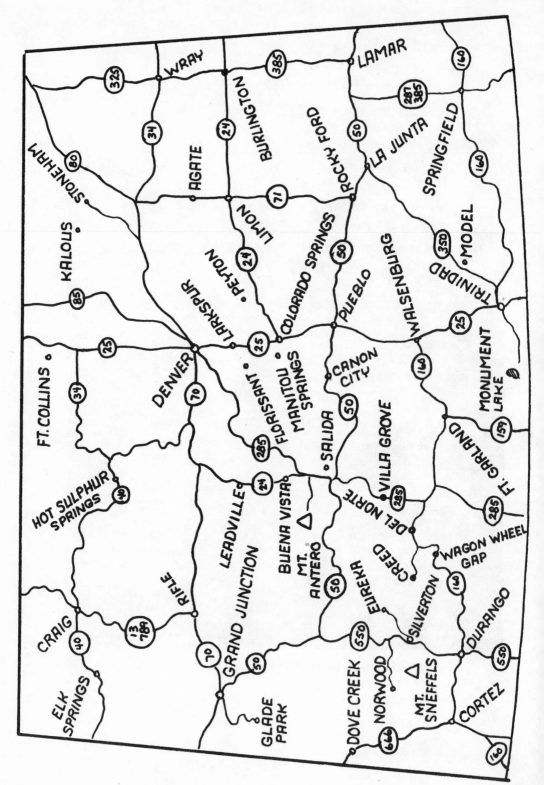

Colorado

Colorado

Bisected from north to south by the Continental Divide of the Rocky Mountains, Colorado has the highest average altitude of any other state. Eastern Colorado is part of the high Great Plains, a rolling, semiarid region that climbs gradually westward into the Rocky Mountain foothills. The Rockies occupy the central and western regions of the state, which is a rugged yet scenic land of forests and meadows. West of the Rockies begins the Colorado Plateau, a rugged arid and semi-arid region of lofty plateaus, steep-sided mesas, and deep canyons.

NOTES OF INTEREST

Near Leadville, Colorado is the largest molybdenum mine in the world.

One of the largest zinc mines in the world is at Gilman, Colorado.

The name Colorado comes from the Spanish, meaning "colored red."

The largest nugget of turquoise on record, weighed nearly 9 pounds, was mined in the King Mine near Manassa, Colorado.

The largest silver nugget, weighing 1840 pounds, was found in the Mollie Gibson Mine at Aspen, Colorado.

Mount Antero, Colorado is considered the highest mineral locality in North America.

The spring and fall are considered the best seasons for collecting in Colorado's desert areas, while summer is suggested for the mountain areas.

The Colorado Bureau of Mines in Denver and the University of Colorado Museum in Boulder both offer notable mineral exhibits.

Chaffee County — Nathrop, area NE of town, apache tears, banded rhyolite, garnet; Mt. Antero area, beryl, topaz, flyorite, aquamarine, smoky quartz; Mt. Taylor, garnet, quartz; Salida, area S of town, travertine.

Elbert County — Agate, area E of town, brown jasper, agate, petrified wood; Elbert, area SE of town, petrified wood; Calhan, Paint Mines, jasper.

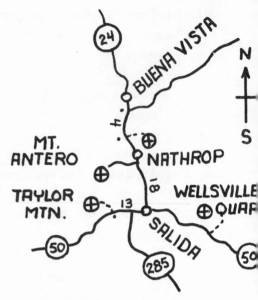

El Paso County — Colorado Springs, mines located SW of city, Topaz, zircon, smoky quartz.

Mineral Occurrences In Colorado

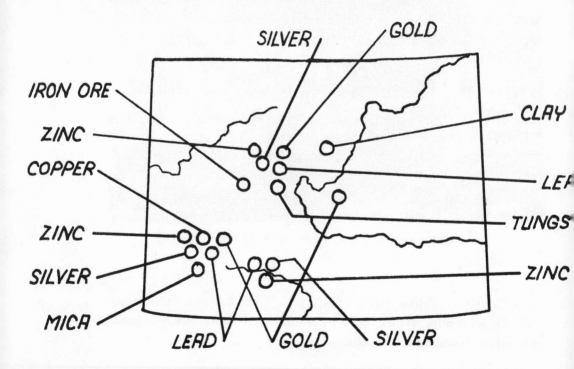

Fremont County — Canon City, area SE of Red Canon Park, Crystal geodes, jasper; Texas Creek, Devil's Hole Mine, N of town, feldspar, mica, rose quartz, beryl.

Grand County — Hot Sulphur Springs, NW on Willow Creek, chalcedony, agate.

Gunnison Countyy — Ohio area mines, topaz, black tourmaline, beryl; Marble, marble quarry, S of town, marble specimens.

Lake County — Leadville, area mine dumps, turquoise.

Diamond drilling, Cripple Creek, Colorado.

Mesa County — Fruita, BookCliffs NE of town, barite crystals; Opal Hill, SW of Fruita, jasper, opalized wood; Glad Park, area S of park, moss agate, desert roses, petrified wood.

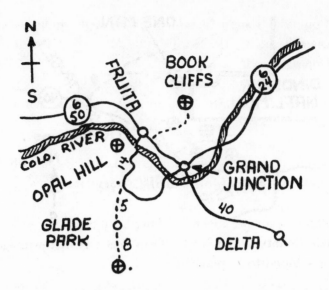

Mineral County — Creed, area crystals, barite, malachite, chalcedony, pyrite, chalcoprite; Wolf Creek Pass, NE of Pagosa Pass, agate, moonstone, geodes.

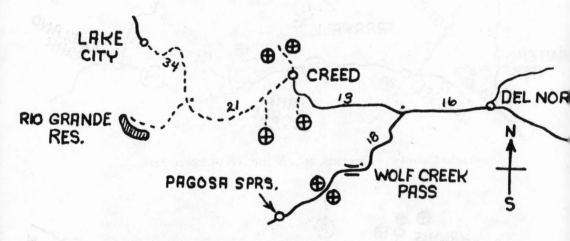

Moffat County — Massodona, area N of town, toward Dinosaur Nat'l. Mon., agate, chalcedony; Elk Springs, area NE of town, agate, jasper; Lone Mountain, N of Elk Springs, jasper.

Park County — Garo, area W of town, agate, chalcedony; Hartsel, area SE of town, agate, jasper; Lake George, area NW of town, fluorite, quartz crystals; Tarryall, Spruce Creek Camp, topaz, feldspar, quartz; Antero Res. area SW along hwy 285, jasper, agate, travertine.

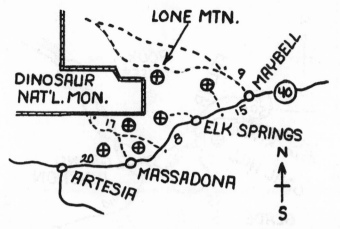

Rio Grande County — Del Norte, Twin Peaks, N of town, agate, opal, thundereggs, bloodstone, petrified wood.

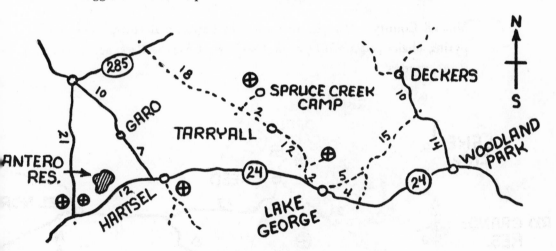

Saguache County — Sargents, area N and NE of town, agate.

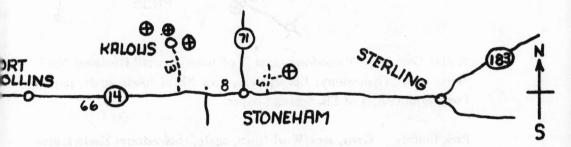

Weld County — Stoneham, area NE of town, barite roses; Kalous, area N and NE of Kalous, jasper; Pawnee Buttes, NW of Kalous, agate, jasper.

73

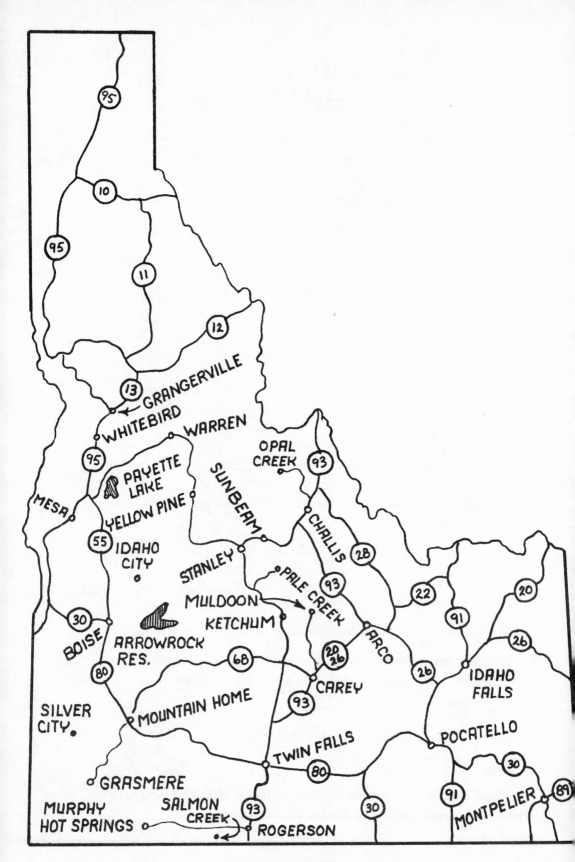

Idaho

Idaho

The oddly-shaped state of Idaho is wholly mountainous, with the main ranges of the Rockies running northwest by southeast along the Montana border. In the south, the wide valley of the Snake River traverses the southern section of the state in a rather sweeping arc, forming part of the western boundary. The Clearwater and Salmon River Mountains in the central region of the state form a barrier between the northern and southern sections. To the west are the Seven Devils Mountains. Bordering the Snake River Plain on the north are the Sawtooth and Pioneer Mountains, the Lost River Range and the Lemhi Range. Southeast of the Snake Valley are the outliners of the Teton and Wasatch Ranges. The Snake River Plain, some 350 miles long, is a region of lava beds, volcanic formations, and sand dunes. It merges in the south and southwest with a desertlike basin.

NOTES OF INTEREST

Idaho's climate offers cold winters and hot summers with an annual rainfall of from 8 to 20 inches and winter snowfall of up to 100 inches in the mountains. Therefore summer and fall are considered the best seasons for collecting.

The Coeur d'Alene district in Idaho is a prominent place for native silver.

The main mining district in Idaho is Shoshone County.

Tetrahedrite is well distributed among the mining districts in Idaho.

In 1865, an enormous mass of crystalline proustite, weighing over 500 pounds, was discovered at the Poorman Mine in Silver City.

Adams County — Bear, area mine dumps, epidot, chrysocolla, garnet, copper specimens.

Benewah County — Fernwood, Crystal Peak, NE of town, quartz crystals; Emerald Creek, S of Fernwood, star garnets.

Blain County — Muldoon, N of Carey, agate, chalcedony, jasper; Goose Creek, S of Carey, chalcedony, agate, jasper; Galena, Pole Creek Summit, crystal geodes.

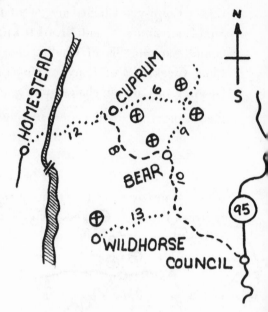

Mineral Occurrences In Idaho

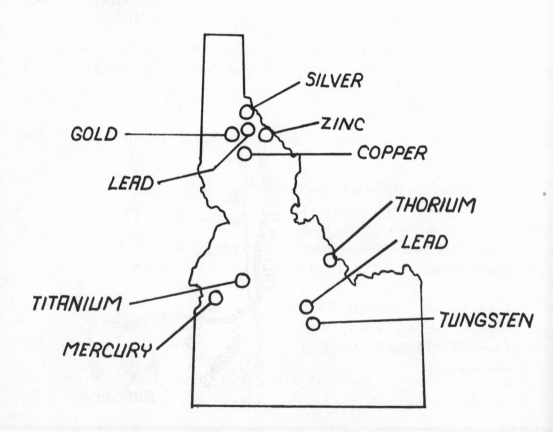

Custer County — Challis, area W of town, red agate, banded opalagate; Sunbeam, mines N and NE of town, carnelian, amethyst geodes; Min. Grand Canyon, SE of Challis, area S of Canyon, agate, chalcedony; Lime Creek, NE of Canyon, chalcedony, agate, agate nodules; Malm Gulch, E of Canyon, chalcedony, agate nodules.

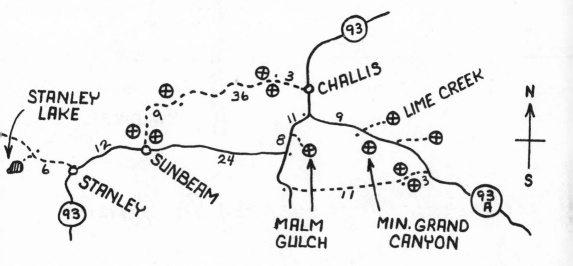

Idaho County — Whitebird, area S along E side of hwy, chalcedony, agate; Salmon River, gravels, agate, garnet; Warren Mine, crystal, quartz, topaz.

Lemhi County — Leesbury, NW of Salmon, petrified wood; Myers Cove, NE of Challis, fluorite, agate, common opal.

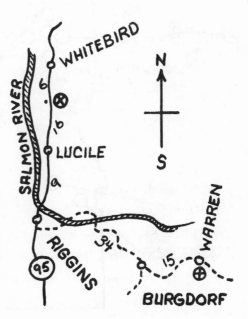

Onyhee County — Silver City, area NW of town, petrified wood, chalcedony; Silver City Mines, barite, fluorite, pyrite, malachite, quartz; Marsing, Poison Creek, SW of town, jasper-agate wood; Squaw Creek, S of Marsing, agate and opal nodules.

Washington County — Midvale, Sage Creek area, W of town, red agate, petrified wood; Manns Creek, NW of Midvale, opalized wood; Crane Creek Res., jasper; Weiser, area W of town near the Snake River, chalcedony, quartz, agate, geodes.

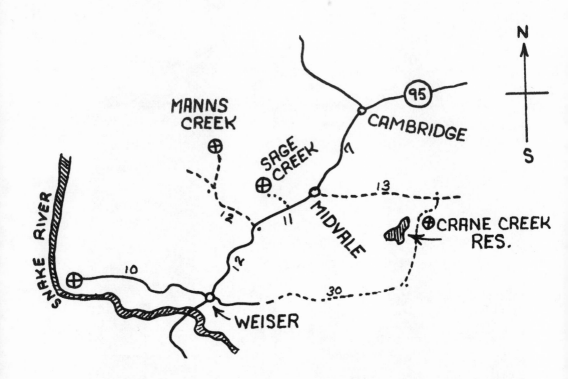

NOTES OF INTEREST

Benawah County in Idaho is a prominent area for garnet, while Latah County, Idaho, offers opal.

Idaho State geological surveys available from the Idaho Bureau of Mines and Geology, University of Idaho, Moscow.

The only crystals of chrysocolla that have ever been found were from Mackay, Idaho.

Niobrara chalk formation in Gove County, Kansas.

Kansas

Kansas

Containing the geographical center of the United States, the state of Kansas descends gradually from 4135 feet on its western border to about 750 feet in the southeast. In the western sections there are high plains, which descend on the east over an irregular slope to the low plains. This region is marked by badland erosion, steep slopes, and isolated buttes. In the eastern section of the state lies a great prairie, characterized by broad valleys and low hill ridges.

NOTES OF INTEREST

State geological surveys are available at the Kansas State Geological Survey, University of Kansas, Lawrence.

The city of Hutchinson grew as a result of the salt discovery in 1887.

Coal is found mainly in the southeast and eastern areas of Kansas, while zinc and lead are found in the extreme southeastern corner of the state.

Petroleum ranks first in the value of mineral resources in Kansas. The chief oil fields are in the central and southeastern sections of the state.

Wallace County — Sharon Springs, area 9 mi. E, S side of hwy, moss opal.

Logan County — Elkader, area N and NW of town, along the Smoky Hill River, moss opal.

Clark County — Ashland, area N of town, moss opal.

Cloud County — Concordia, area NW of town, along the Republic River, agate, jasper.

Niobrara chalk overlaying Carlyle Shale, Logan County, Kansas.

Osborne County — Downs, Great Spirit Springs, SE of town, banded travertine.

Ottawa County — Minneapolis, area NW of town, petrified wood.

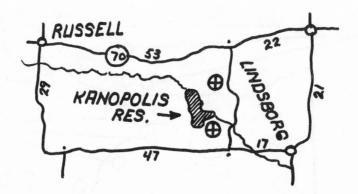

Ellsworth County — Kanopolis Reservoir, NW of Lindsborg, area S of reservoir, celeslite crystals; Saline Quarry, NE of reservoir, barite roses.

A Dakota formation, a combination of sandstone and shale, located around Ellsworth County, Kansas.

Mineral Occurrences In Kansas

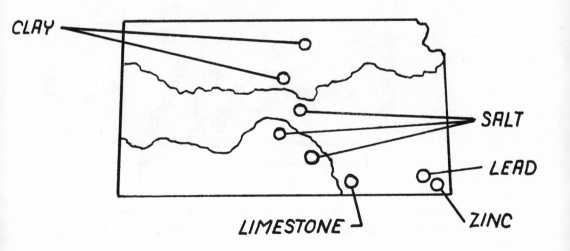

Niobrara chalk formation, Gove County, Kansas.

Monument Rocks, Niobrara chalk formations, Gove County, Kansas.

The crystal form of celestite, located in Brown County, Kansas.

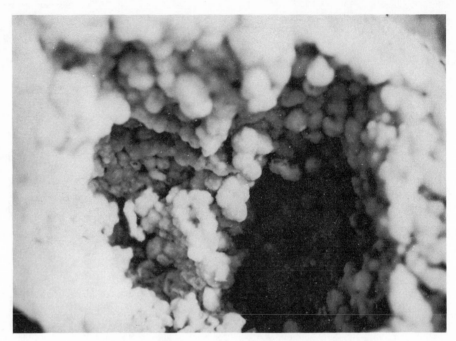

Calcite in geode, located in Butler County, Kansas.

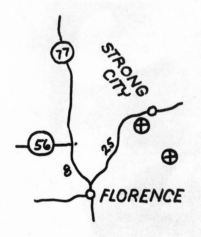

Marion County — Florence, area NE of town, along S side of the hwy, geodes; area S of Strong City, petrified wood.

Barber County — Medicine Lodge, area SE of town, agate-jasper, petrified wood.

Cherokee County — Galena, area mines, chalcopyrite, marcasite.

Galena (lead), located in Cherokee County, Kansas.

Galena (lead), located in Cherokee County, Kansas.

Alternating limestone and shale located at the spillway of Tuttlecreek Dam in Pottawatomie County, Kansas. Note small fault in strata.

NOTES OF INTEREST

Temperatures are normally extreme in Kansas during the summer and winter months, therefore late spring and fall are considered the best seasons for collecting.

The University of Kansas Museum in Lawrence has a notable mineral exhibit.

Limestone deposits are in the vicinity of Kansas City.

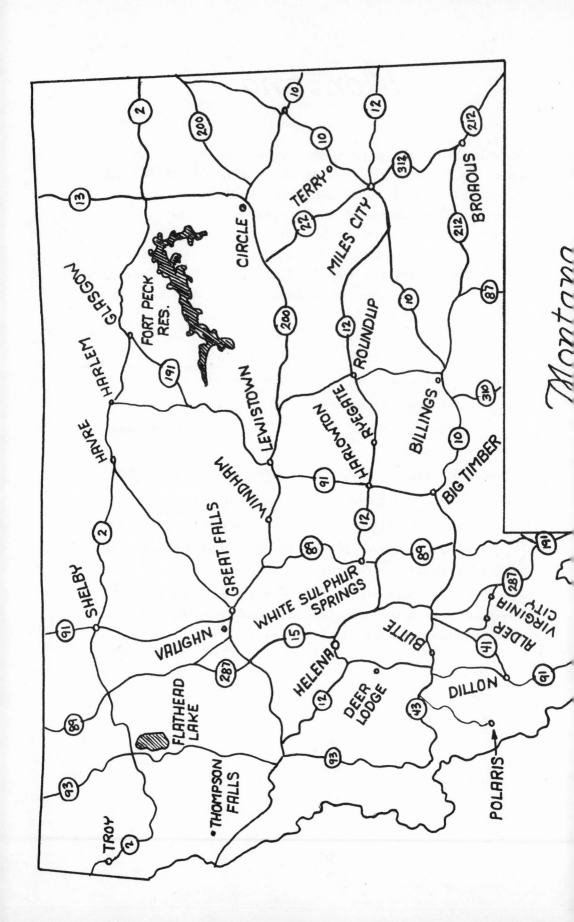

Montana

Montana lies in two physiographic regions, the Northern Rocky Mountains in the west, which occupy about 40 percent of the total area, and the Great Plains to the east. The Rocky Mountains, consisting of several broken ranges of metamorphic rocks, run generally northwest-southeast in parallel ridges. The main ones are the Lewis Range, the Cabinet Mountains, the Big Belt Mountains, and the Bitterroot Range, which lies along the Idaho border. To the extreme south is the Absaroka Range, while through central Montana to the Canadian border are the detached outlines of the Rockies.

NOTES OF INTEREST

Montana is the nation's main source of gem sapphires.

Montana leads all other states in the production of manganese.

Montana gets its name from the Spanish word meaning "mountainous country."

Elkhorn, Montana, is a locality for fine specimens of hemimorphite.

Butte, Montana, has furnished good specimens of sphalerite.

The Montana School of Mines in Butte has a notable mineral exhibit.

When gold was discovered at Alder Gulch in 1863, 100 dollars of gold a day for one man was considered a poor average.

For 250 miles, from Billings to the North Dakota border, the Yellowstone River gravel bars and beachlands offer the collector gemstone wealth in the form of agate and jasper.

Beaverhead County — Dillon, area SE of town, agate; Elkhorn Springs, area N along W side of the hwy, quartz crystals.

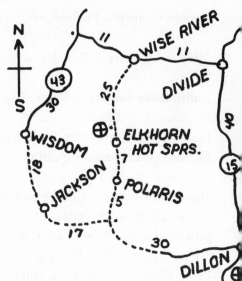

Big Horn County — Dryhead, area E at Big Horn Canyon, along both sides of the Big Horn River, geodes, agate nodules.

Cascade County — Vaughn, areas NW and SW of town, black agate wood.

Custer County — Miles City, area N of town, along the Yellowstone River to Terry, moss agate.

Deer Lodge County — Anaconda, Lost Creek area, 3 mi. N of town, black agate wood.

Gallatin County — Bozeman, Gallatin Range SE of town, petrified wood; Maudlow, Horseshoe Hills, N of town, moss agate.

Mineral Occurrences In Montana

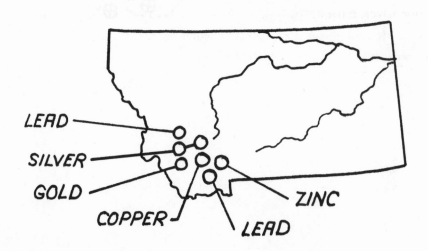

Granite County — Phillipsburg, area mine dumps, rhodochrosite.

Jefferson County — Boulder, area mines, amethyst, quartz crystals.

Judith Basin County — Utica, Yogo Gulch, sapphire mines, sapphire.

Lewis And Clark County — Helena, areas W and SW of city, S of hwy 12, rose quartz, smoky quartz.

Madison County — Alder, area E of town, along both sides of hwy, garnet; Twin Bridges, area W of town, quartz crystals; area N of Twin Bridges, along both sides of hwy 41, jasper; area SE of the Ruby Reservoir, garnet; area SW of reservoir, wonderstone.

Park County — Carbella, Tom Miner Basin area SW of town, jasper, petrified wood; Gallatin Range, NW of Carbella, petrified wood.

Yellowstone County — Custer, areas SW and NE of town, along the Yellowstone River, moss agate.

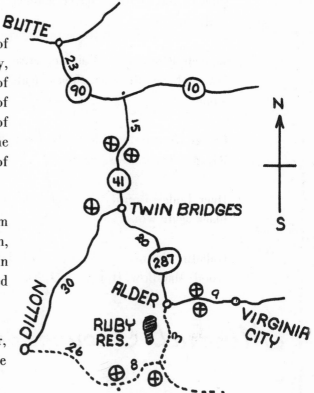

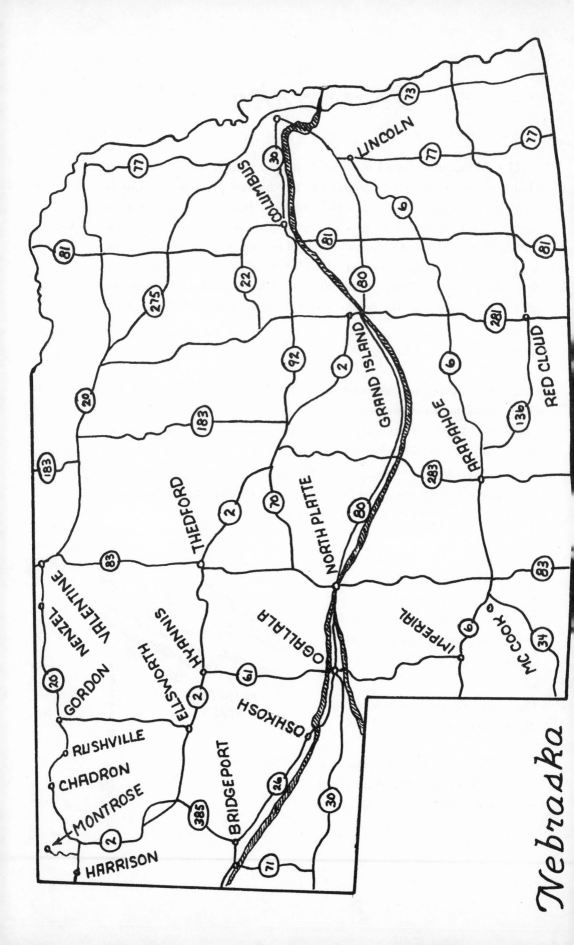

Nebraska

The rectangular-shaped state of Nebraska, which lies in the Great Plains, slopes from west to southeast. The highest point in the state is 5340 feet, located near the state's western boundary. The semiarid western part of the state is a dissected plain which rises to broken tableland in the west and includes a small section of the badlands in the northwest. The eastern section of the state is undissected plain with drift hills to the southeast. Sand hills, somewhat stablized by the prairie grass, extend through the north-central and central regions of the state.

NOTES OF INTEREST

Diggings along the Niobrora River in Cherry, Daws, Sheridan, and Sioux counties have produced many fossil remains of mid-Pleistocene mammals. Agate, Bridgeport, Hays Springs, Sidney, and Valentine are said to be good localities for the fossil collector.

State geological surveys are available from the Nebraska Conservation and Survey Division, University of Nebraska, Lincoln.

Cherry County — Valentine, areas SE and SW of town, along Niobara River, petrified wood; Merriman, area S of town, along S side of Niobara River, petrified wood.

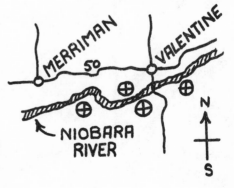

Cheyenne County — Sidney, area SE of town, agatized wood.

Dawes County — Chadron, area NW of town, W of hwy 385, fluorescent chalcedony. Area N of town, agate, jasper.

Deuel County — Chappell, area W of town, along S side of the hwy, agatized wood.

Garden County — Lewellen, area SW of town, along S side of the Platte River, petrified wood.

Morrill County — Bayard, areas SW and E of town, along North Platte River, agate, jasper; Bridgeport, area E of town, along S side of the Platte River, petrified wood.

ᴹineraℓ Occurrences In Nebraska

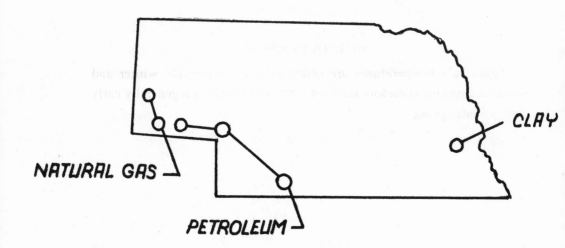

Scotts Bluff County — Scottsbluff, area 8 mi. N of town, along E side of hwy 71, opalized wood; area SW of Scottsbluff, along the North Platte River, agate, moss agate, carnelian, quartz.

Sioux County — Montrose, NE of Harrison, area N and E of Montrose, chalcedony, jasper, agate-wood.

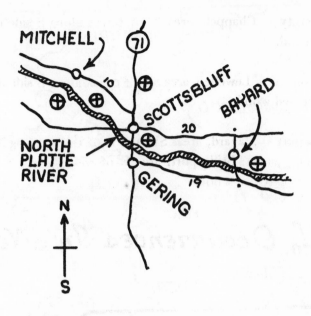

NOTES OF INTEREST

Nebraska's temperatures are often extreme during the winter and summer months, therefore the best time for collecting is generally early fall and late spring.

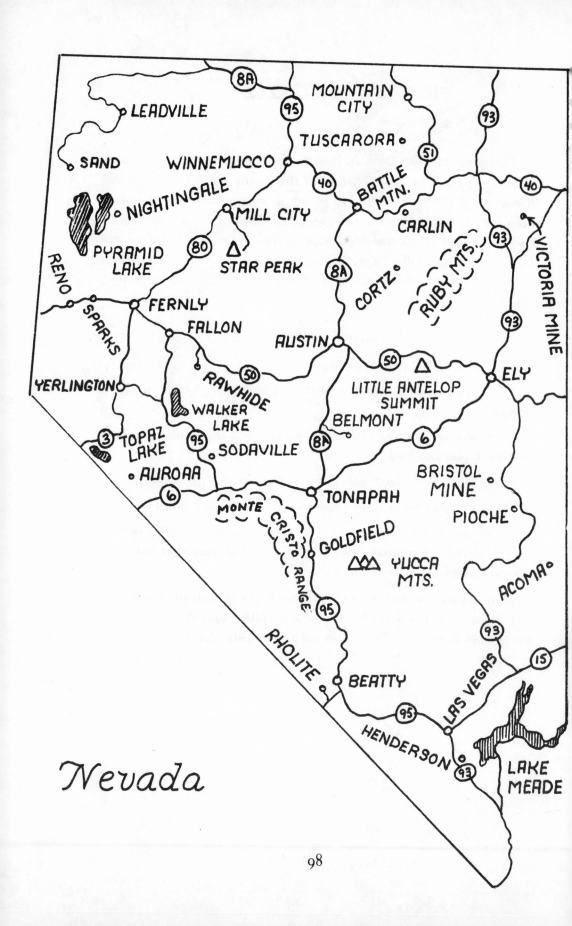

Nevada

Nevada

The semitriangular state of Nevada lies within the arid Great Basin region, cut off from the moisture of the Pacific by the Sierra Nevada Range on the west, which also serves as a part of the northwestern boundary between Nevada and California. The region is characterized by its great aridity, sparse population, extensive mining operations, and its broad plains clothed only in sagebrush. Hundreds of mines, small and large, dot the entire state and offer many colorful and interesting minerals.

NOTES OF INTEREST

Rhodochrosite is a common mineral in the silver mines at Austin.

One of the most valuable minerals found in Nevada today is uranium.

Chief mining centers in Nevada today are Ely, Pioche, Tonopah, Goldfield, and Lovelock.

There is a notably large open-pit copper mine at Ruth, Nevada.

Nevada gets its name from a Spanish word meaning "snow-clad."

Spring and fall are considered the best seasons for collecting in most northern areas of Nevada, while spring, fall, and winter are suggested for most southern areas of the state.

State geological surveys are available at the Bureau of Mines, Rare and Precious Metals Experiment Station, 1605 Evans Ave., Reno, also the Nevada Bureau of Mines, University of Nevada, Reno.

Churchhill County — Fallon,
Lahontan Dam, area S of Dam,
wonderstone; NE shore of reservoir,
agate, jasper; Wonderstone
Mountain, E of Fallon, ryholite,
wonderstone; area S of mountain,
lace agate; Brady Hot Springs, area
N, W side of hwy 40, agate, jasper.

Clark County — Goodsprings, SW
of LasVegas, area mines, smithsonite, azurite, malachite, garnet, pyrite,
cinnabar, feldspar crystals; Gyp Cave, area E of cave, agate, chalcedony;
Henderson, area S of town, chalcedony, agate, Black Cliffs, SE of
Crystal, colemanite, opal-agate; Hoover Dam, area SE of dam, S side of
hwy 93, agate.

Sampling high-grade outcrop of mercury ore. Pershing Mercury,
Lovelock, Nevada.

Mineral Occurrences In Nevada

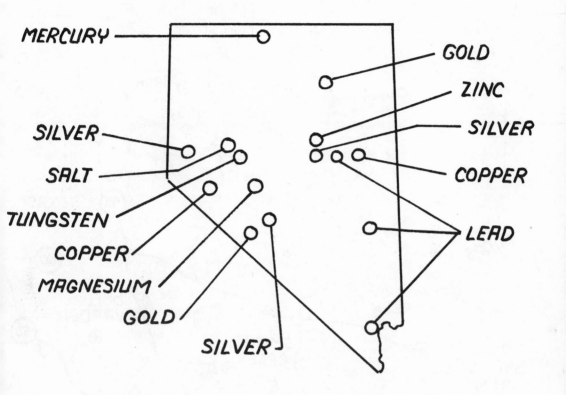

MERCURY

GOLD

ZINC

SILVER

SILVER

SALT

COPPER

TUNGSTEN

COPPER

MAGNESIUM

LEAD

GOLD

SILVER

Elko County—Midas, Rock Creek area, S of town, agate, chalcedony, opalite wood; Ruby Valley, Ruby Mountains, beryl, mica, garnet; Smith Creek, N of Ruby Valley, quartz crystals; the Dolly Varden area, jasper, chalcedony, chrysoprase; Jarbridge, Jack Creek area mines, crystal specimens; Jackpot, Texas Spring, petrified wood.

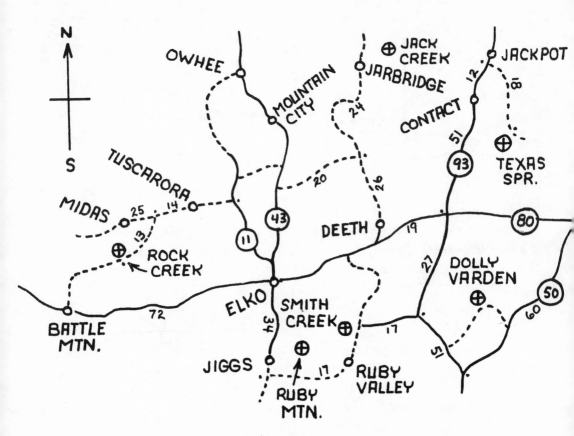

Esmeralda County—Coaldale, area E along hwy 95, jasper, agatewood; area S of Coaldale, apache tears; Goldfield, Goldfield Gem Claim, chalcedony, apache tears, agate, opalite; area SE of Goldfield, E of hwy 95, opalite, sulphur crystals, chalcedony; Gold Point, agate-wood; Chandelaria, mines NW of Coaldale, turquoise, copper ores, scheelite.

Humboldt County — Paradise Valley, Firestone Opal Mine, fire opal; Denio, area W of town, agate; Agate Peak, SE of Denio, chalcedony, sulphur mine S of town, sulphur crystals, gypsum, cinnabar; Black Rock Desert, petrified wood, jasper, opal-agate geodes; Virgin Valley, area mines, fire opal, opal-wood, agate, oarnelian.

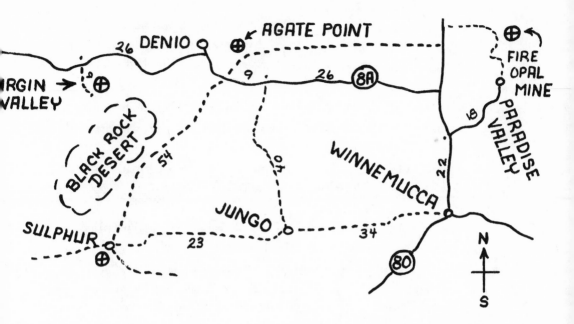

Lander County — Battle Mountain, area mines, turquoise, pyrite, azurite, malachite, chrysocolla, cuprite; Dacie Creek, SW of Battle Mountain, petrified wood; Buffalo Valley, titanium, fluorspar, rhodochrosite.

Lyon County — Fernly, area SW of town, agate, jasper; Fort Churchill State Park, areas S & SE of Park, agate, gypsum, calcite, jasper; Yerington, area E of town, jasperwood; Singatse Range, SW of Yerington, malachite, garnet, azurite, chalcopyrite; Wilson Canyon, NE of Wellington, petrified wood; Ferris Can., fluorite.

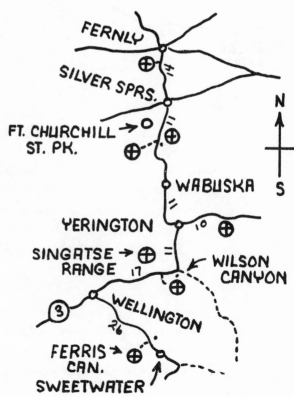

Mineral County — Aurora, quartz crystals, chalcedony, chalcopyrite; Mina, area SW of town, jasper; Montgomery Pass, area W of Pass, obsidian nodules; area S of Pass, geodes.

Nye County — Manahattan, area mines, calcite; Beatty, area NE of town, chalcedony, geodes; Duckwater, area SW of town, agate, geodes.

Pershing County — Lovelock, area NW and W of town, wonderstone, obsidian, opalite; Star Peak, agate geodes.

Washoe County — Sano, N of Pyramid Lake, petrified wood, chalcedony; Leadville, zinc, lead ores; Nightingale area mines, calcite, garnet, scheelite; Duck Lake, area NW of lake, jasper; Tuledad Canyon, W of lake, chalcedony, agate; Petrified Stump State Park, area NW of park, along hwy 34, opal, apache tears, agate-wood.

White Pine County — Connors Pass, SW of Ely, calcite; Ely, area NW of town, garnet, spessartite.

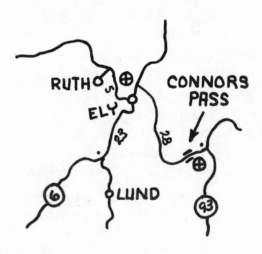

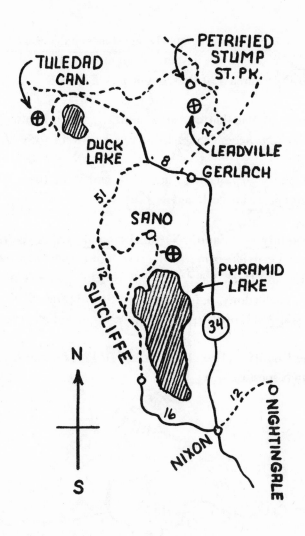

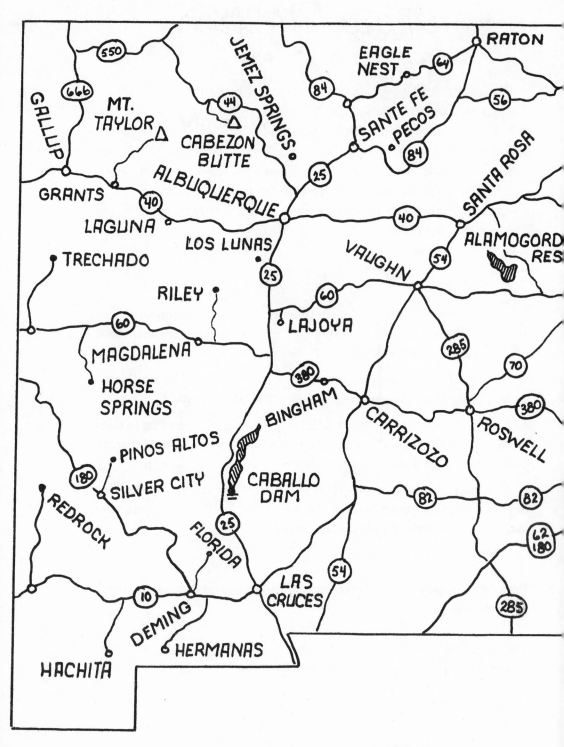

New Mexico

New Mexico

The state of New Mexico offers many features of forested mountains, plateaus, plains, and deserts. The northwestern section of the state lies on the Colorado Plateau, which is marked by lava flows, volcanic cones, colorful buttes, and mesas. The southwestern and southern regions are characterized by eroded black mountains and desert plains. To the east the state falls within the Great Plains, and is dissected in the north by canyons and mesas. Farther south lies the Pecos Valley bordered on the east by sharp slopes, marking the west limit of the extremely level Staked Plain.

NOTES OF INTEREST

The University Of New Mexico Geology Museum, Albuquerque, and the New Mexico Institute of Mining and Technology, Socorro, both offer notable mineral **exhibits**.

State geological surveys are available from the New Mexico Bureau of Mines and Mineral Resources, Camous Station, Socorro.

Spring and fall are considered the best seasons for collecting in most areas of New Mexico, although summer is the best for mountain areas. Collecting can be done nearly year around along the Rio Grande.

McKinley County, New Mexico, is said to be an outstanding collecting site for garnet, while Sante Fe, Grand, and Otero counties are said to be good collecting areas for turquoise.

Bernalillo County — Rio Purco area, NW of Albuquerque, agate, jasper, petrified wood.

Chaves County — Roswell, Pecos River gravels, NE of town, amber quartz; area NE of Roswell, agate, petrified wood.

Catron County — Quemado, area NE of town, agatewood; Horse Springs, agatized wood; area SE of Horse Springs, quartz crystals, agate; Mogollon, area NE of town, chalcedony roses; Luna, area NW of town , amethyst geodes, agate; area SE of Luna, amethyst, blue agate; Apache Creek, Turkey Park, agate, quartz crystals.

Strange formations from part of City of Rocks State Park, near Silver City, in southwestern New Mexico. This is a popular picnic area for residents and visitors. *Courtesy New Mexico State Tourist Bureau.*

Mineral Occurrences
In New Mexico

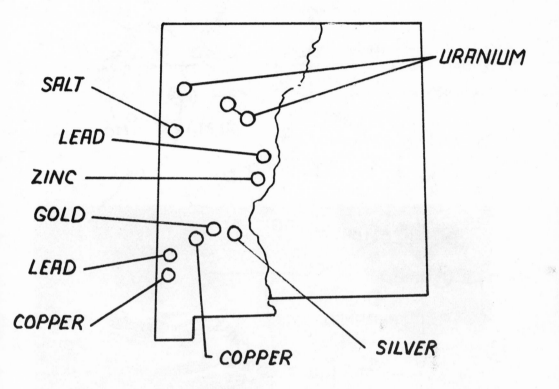

Dona Ana County — Hatch, diversion dam, S of town, moss opal; Rincon, area NE of town, agate, thundereggs, jasper.

Grant County — Mule Creek, area W of town, obsidian nodules; Redrock, NE in Ricolite Gulch, ricolite; Tyrone, Azure Mine, turquoise, apatite, feldspar; Pinos Altos, area N of town, moss jasper, carnelian, chalcedony roses; Mimbrea, Rabb Canyon, NE of town, moon stone; Hachita, area W, SW, and S of town, moss agate, moonstone, jasper.

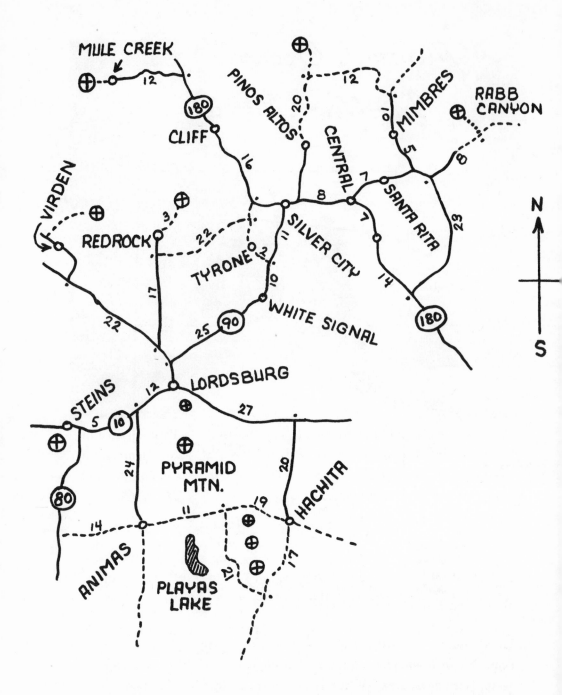

Hidalgo County — Lordsbury, mine dumps, S of town, bornite, azurite, lead; Pyramid Mountain, S of Lordsburg, chalcedony; Animas, Playas Lake, blue chalcedony, agate, jasper; Steins, area S of town, quartz, chalcedony; Virden, area NE of town, petrified wood.

These massive monoliths are in City of Rocks State Park, an area of huge, oddly shaped stones between Silver City and Deming, in southwestern New Mexico. City of rocks is a popular camping ground. *Courtesy New Mexico State Tourist Bureau.*

Lincoln County — Ancho, area N of Town, jasper; Corona, Gallinas Mountains, E of town, barite crystals.

Luna County — Nutt, area N of town, desert roses; Florida, Cooks Peak, NE of town, chert, quartz; Fluoride Ridge, NE of Deming, jasper, agate; Hermanas, area NW and NE of town, agate; Columbus, area W of town, Mexican onyx; Baker Ranch, NE of Columbus, agate, thunder-eggs, crystal geodes.

McKinley County — McGaffey, Zuni Mountains, petrified wood; Blue-water Lake, area NE of the lake, agate, petrified wood.

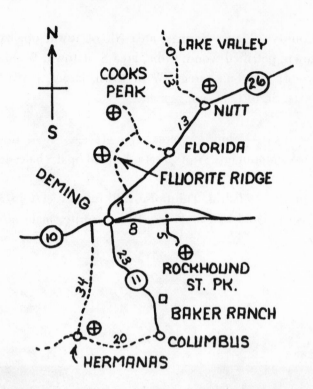

Rio Arriba County — Abiquiu, area mines, malachite, azurite; Dixon, area mines, garnet, apatite, quartz, beryl, tourmaline; Ojo Caliente, area mine dumps, beryl, garnet, kyanite.

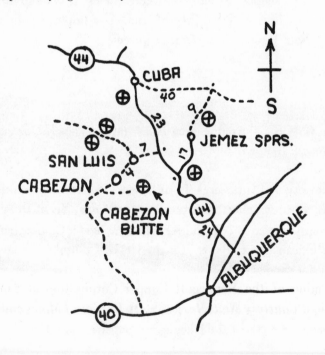

Sandoval County — Jenes Springs, area NE of town, opalized wood; area S of town, petrified wood; Cuba, area S of town, W side of hwy, petrified wood; San Luis, area NW of town, jasper, petrified wood; Cabezon Butte, SE of Cabezon, geodes.

San Juan County — Shiprock, area river gravels, SE of town, agate; Crystal, Chuska Mountains, N of Crysta, banded opal, chalcedony.

Sante Fe County — Madrid, Turquoise Hill, turquoise; Galisted, area N of town, jasper; Golden, San Pedro Mine, azurite, malachite, garnet, chrysocolla.

Open-pit mine of the Kennecott Copper Corporation at Santa Rita, New Mexico. *Courtesy New Mexico State Bureau of Mines and Mineral Resources.*

Open-pit mine of the Molybdenum Corporation of America at Questa
New Mexico. *Courtesy New Mexico State Bureau of Mines and Mineral
Resources.*

Sierra County — Elephant Butte, opalized wood; Chloride, Montezuma
Mine, amethyst; Diamond Creek, NW of Chloride, topaz; Engle, area SE
of town, chalcedony, opalized wood; Cutter, area SW of town, carnelian; Caballo Dam, area NE of Dam, picture rock; Hillsboro, area N
and NW of town, opal agate, chalcedony roses; Kingston, Comstock &
Lady Franklin Mines, rhodonite, rhodochrosite; Lake Valley, areas N
and NE of town, agate.

Socorro County — Riley, area SW of town, palm and picture wood;
Ladron Mountains, NE of Riley, amethyst, quartz; Kelly, area mines,
aurichalcite, azurite, malachite, barite, smithsonite; Bingham, area
mines, quartz crystals, fluorite, barite, calcite, onyx.

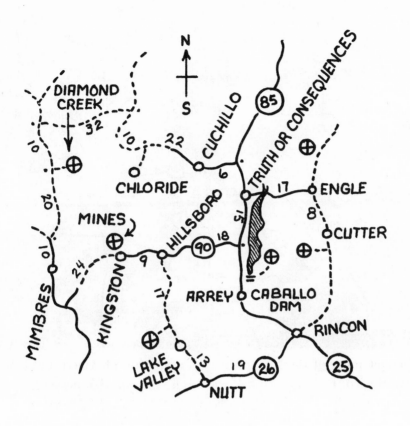

Valencia County — Grants, area NE of town, apache tears; Laguna,
agate-wood.

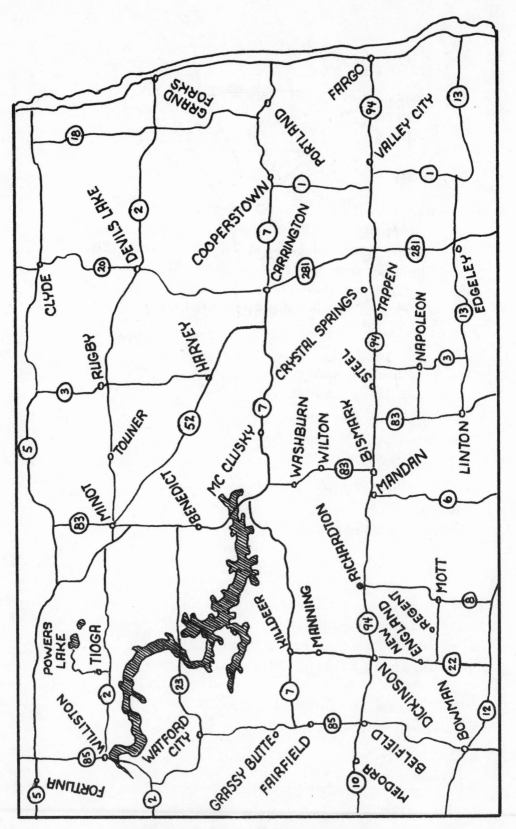

North Dakota

North Dakota

The state of North Dakota is the geographical center of North America. Its surface is generally flat, underlain by layers of sandstone and shale which contain extensive veins of lignite and beds of bentonite. The state rises in three broad steps from the east to west and includes sections of the central lowlands to the east and Great Plains in the west. In the eastern section of the state is the Valley of the Red River of the north which was once the floor of a glacial lake. Extending south from Manitoba through central North Dakota is the Drift Prairie, an expanse of glacial drift. Missouri Plateau lies in the western section of the state, extending from the Missouri escarpment to the Rocky Mountains.

NOTES OF INTEREST

Lignite is abundant in western regions of North Dakota and is so close to the surface that strip mining is possible.

Large deposits of clay, suitable for pottery, and bentonite are found in the southwestern regions of the state.

The Yellowstone and Missouri River gravels in Williams and McKenzie counties are popular collecting localities for agate.

State geological surveys are available from the North Dakota Geological Survey, University Station, in Grand Forks.

McKenzie County — Cartwright, area N of town, along S side of the Missouri River, moss agate; Watford City, area SE of town, along the N side of the Little Missouri River, Theodore Roosevelt Nat. Mem. Park, North Unit, area SW along Little Missouri River, agate-wood.

Slope County — Marmarh, area N of town, along E side of the Little Missouri River, agate-wood.

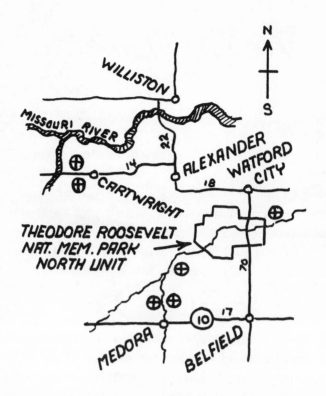

A small abandoned dozer and scraper operation. Old pit (left center, spoil piles (right center). Located at Pederson Mine, McKenzie County, North Dakota. Abandoned in 1953. *Courtesy North Dakota Geological Survey, University of North Dakota.*

A view of a dozer and scraper operation taken from the spoil pile. Dozer and loading shovel in active pit area. Location: Ecklund Mine, Burleigh County, North Dakota. *Courtesy North Dakota Geological Survey, University of North Dakota.*

Spoil piles from an abandoned dragline operation as viewed from the roadway. Location: Baukol, Noonan Pit Divide County, North Dakota. *Courtesy North Dakota Geological Survey, University of North Dakota.*

Hettinger County – New England, large area NW of town, along both sides of the river, petrified wood; Regent, areas NW and E of town, along both sides of the river, petrified wood; Mott, area E and W of town, along both sides of the river, silicified wood.

Mineral Occurrences In North Dakota

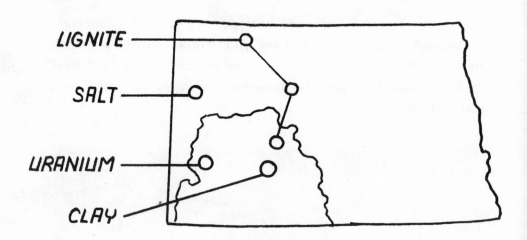

Stork County – Richardton, areas S of town, along river to Lake Tschida, jasper, chalcedony.

Kidder County – Crystal Springs, area W and S of town, S side of hwy 94, quartz gems.

Ward County – Minot, gravel pits, SW of town, quartz minerals.

Mountrail County – White Lake, NW of Stanley, area W of lake, halite, glauberite.

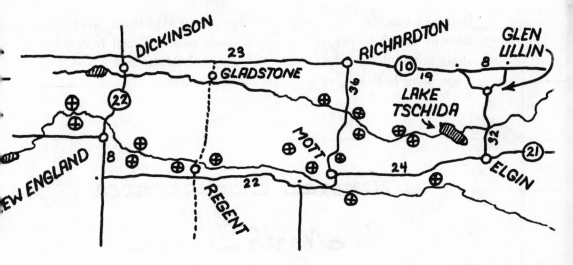

NOTES OF INTEREST

Although the summer is short and hot in North Dakota, it is the best time for collecting.

Sodium sulphate is produced from dry lake beds in the northwestern part of the state.

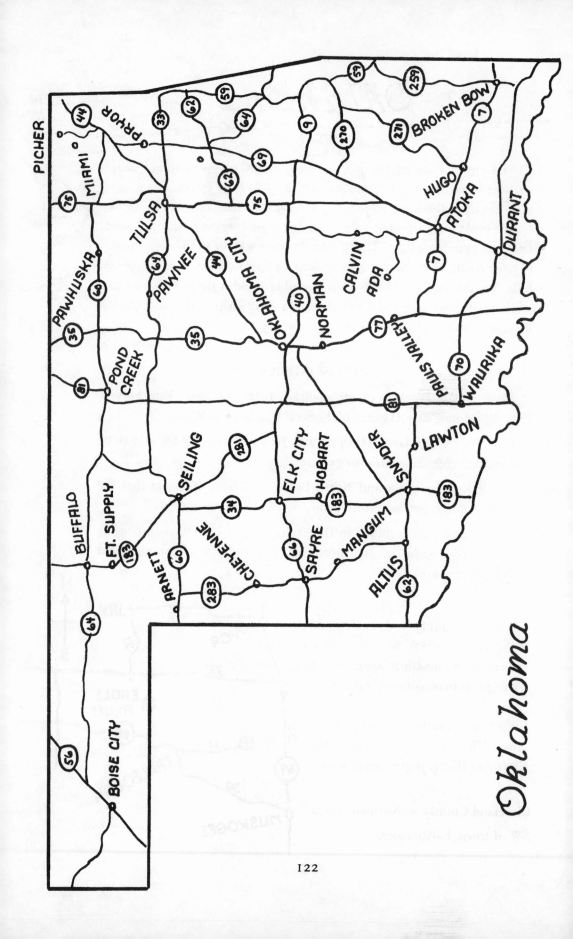

Oklahoma

Oklahoma

The entire state of Oklahoma is essentially plains country, sloping southeast and interrupted occasionally by low hilly tracts. The western counties lie in the level, almost treeless plains, which descend in the east over a broken escarpment to a dissected plains region containing saline flats and gypsum ledges. This gradually gives way to an extensive prairie of underlying shale formations broken by low sandstone and limestone hills. To the east are the Boston Mountains and a rugged plateau area while to the extreme southeast there is a small strip of the Gulf Coastal Plain with soil of clay and sand.

NOTES OF INTEREST

State geological surveys are available from the Geological Survey, Regional Mining Supervisor, 205 Federal Building, in Miami.

Beckham, Blain, Greer, Harper, and Jackson counties are said to be outstanding collecting sites for gypsum.

Along the Cimarron and North Canadian Rivers, it is said that the gravel bars often produce fossil bones of Pleistocene mammals.

Bituminous coal is mined in the eastern regions of Oklahoma, while the tri-state region in the northeast corner is a major zinc and lead-producing area.

Cherokee County — Tahlequah, Hubbert Quarry, W of town, travertine; Eagle Bluff area, NE of Tahlequah, marcasite crystals.

Cimarron County — Boise City, area NW of town, near the Cimarron River, jasper, agate wood.

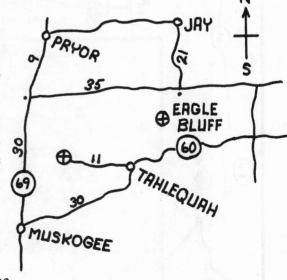

Cleveland County — Norman, area SW of town, barite roses.

Comanche County — Lawton, area 20 mi. W of town, S of hwy 62, barite balls; area NW of Lawton, quartz crystals.

Coton County — Taylor, Cherry Canyon, NE of town, calcite and barite crystals.

Headframes (derrick) used to mine zinc-lead ore in the Tri-State field at Picher, Oklahoma. *Courtesy United States Department of the Interior.*

Mineral Occurrences
In Oklahoma

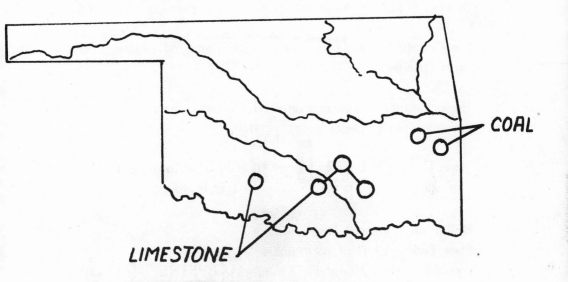

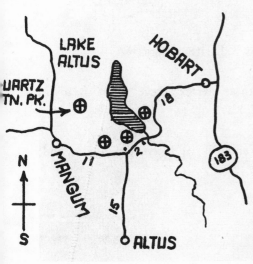

Dewey County — Seiling, area E of town, agate; area SW of Seiling, near the Canadian River, petrified wood, agate.

Greer County — Mangum, Quartz Mountain Peak, NE of town, quartz crystals; Lake Altus, areas E and SW of the Lake, crystal geodes, quartz crystals.

Harper County — Buffalo, area NW of town, geodes; area SW of Buffalo, S side of hwy, agate.

Hughes County — Calvin, area S of town, E side of hwy, jasper, chert, petrified wood.

Jackson County — Altus, area S of town, jasper, agate, petrified wood.

Kay County — Ponca City, Arkansas River area, NE of city, geodes with barite, calcite.

Mayes County — Spavinaw area NE of town, chert nodules; area S of Spavinaw, pyrite, jasper, quartz crystals.

Major County — Fairview, areas NE and NW of town, agate; area SW of Fairview, S side of hwy 60, chalcedony, agate wood.

McCurtain County — Beavers Bend State Park, area N of park, quartz crystals; Crystal Mountain, NW of Beavers Bend Park, crystal; area SW of Crystal Mountain, N of Glover, quartz crystals.

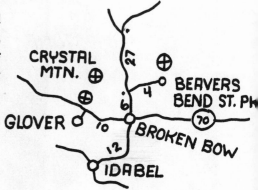

Murray County — Davis, area E of town, barite dollars.

Pawnee County — **Pawnee**, Arkansas River area, NW of town, chalcocite, **malachite**.

Pontotoc County — Ada, areas S and SW of town, petrified wood; mining claim SW of Ada, marcasite.

Headframe (derrick) used to mine zinc-lead ore in the Tri-State field at Picher, Oklahoma. *Courtesy United States Department of the Interior.*

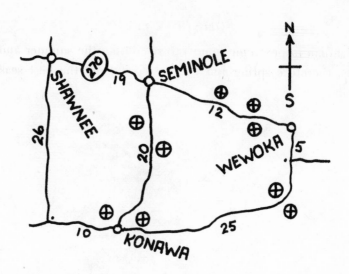

Seminole County — Seminole, area S of town, agate, jasper, chert, petrified wood; Konawa, areas N and NE of town, jasper, chert petrified wood, agate; Wewoka, large area W of town, agate, jasper, petrified wood.

Tillman County — Grandfield, area NE of town, geodes, barite roses.

Headframes (derricks) used to mine zinc-lead ore in the Tri-State field at Picher, Oklahoma. *Courtesy United States Department of the Interior.*

NOTES OF INTEREST

Oklahoma has extreme temperatures during the summer and winter months, therefore spring and fall are considered the best seasons for collecting.

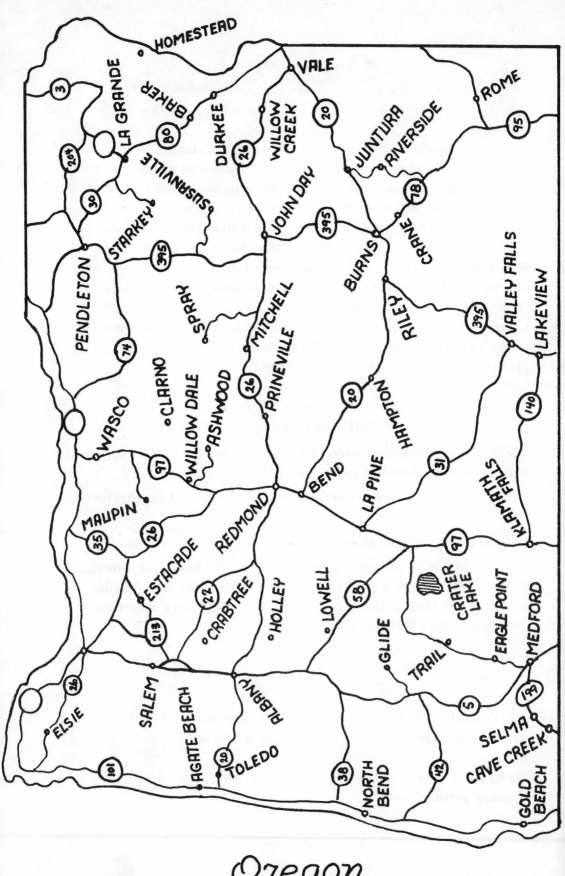

Oregon

Oregon

Running parallel to Oregon's coast, between the Columbia River and the California border, are low rolling hills of the coastal ranges. About 100 miles inland is the Cascade Range, which running north to south, divides the state into contrasting east and west sections. The Cascades section of the state is one of rugged grandeur with its glaciers and lakes. Between the Coast and Cascade Ranges is the fertile trough of the Willamette River Valley, that extends about 170 miles south. South of the valley, the Coast and Cascade Ranges merge in the Klamath Mountains, a dissected plateau. Oregons eastern portion lies in the Columbia Plateau, a broad expanse underlain by horizontal lava beds. In the Columbia Basin, lying in the north of the state, are the rolling plains of the Inland Empire.

NOTES OF INTEREST

Central Oregon offers several thousand square miles of territory where almost every type of quartz may be found.

Lake County, Oregon, is said to be a good collecting locality for obsidian.

Spring, summer, and fall are the best collecting seasons in Oregon.

State geological surveys are available from the Bureau of Mines, Northwest Electrodevelopment Experiment Station, P.O. 492, Albany, also from the Oregon Department of Geology and Mineral Industries, 1065 State Office Building, in Portland.

Baker County — Durkee, area SW of town, jasper-agate; Whitney, area NE of town, agate; Cornucopia, area mines, quartz and quartz crystals.

Clackamas County — Estacada, Clackamas River area, bloodstone, cinnabar, petrified wood.

Columbia County — Vernonia, area SW of town, carnelian, agate, jasper.

Crook County — Eagle Rock, SE of Prineville, agate; Post, area E of town, gem agate, geodes, thundereggs.

Curry County — Gold Beach area, agatized coral, garnet, agate-jasper.

A teenage "rockhound" shows colorful slices of agate from a "thunderegg" found near Prineville, Oregon. The egg-shaped, lava-covered agate nodules normally are not this large, but thousnads two to three inches in diameter are found yearly in the mountains of central and eastern Oregon. Agates, opals, garnets, and other semiprecious stones are found in both coastal and mountainous areas of the state. *Courtesy Oregon State Highway Department Photo 7257.*

Succor Creek wends its way through the rugged canyon it has cut through the centuries in the rock of the Owyhee Mountains of southeastern Oregon. Spectacular scenes such as this are repeated many times through the canyon, which is reached by paved and gravelled roads from Ontario or Jordon Valley. *Courtesy Oregon State Highway Department Photo 6979.*

Mineral Occurrences In Oregon

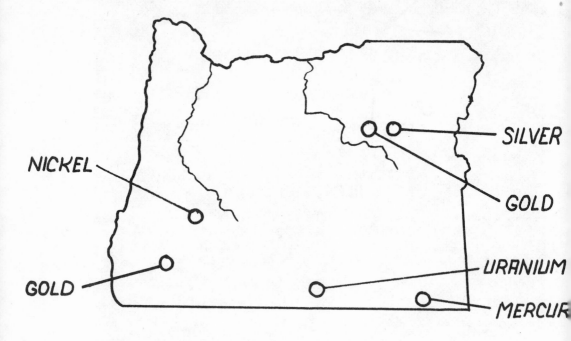

Deschutes County — Brothers, areas N and NE of town, chalcedony, jasper, petrified wood.

Douglas County — Glide, areas E and SE of town, agate, garnet, jasperized wood.

Harney County — Andrews, Alvord Ranch, NW of town, moss agate, jasper, thundereggs; Crane, Buchanan Ranch, N of town, agate, thundereggs; Stinkinwater Mountain, agate, petrified wood; Warm Springs Res., W shore, agate; area W of reservoir, petrified wood.

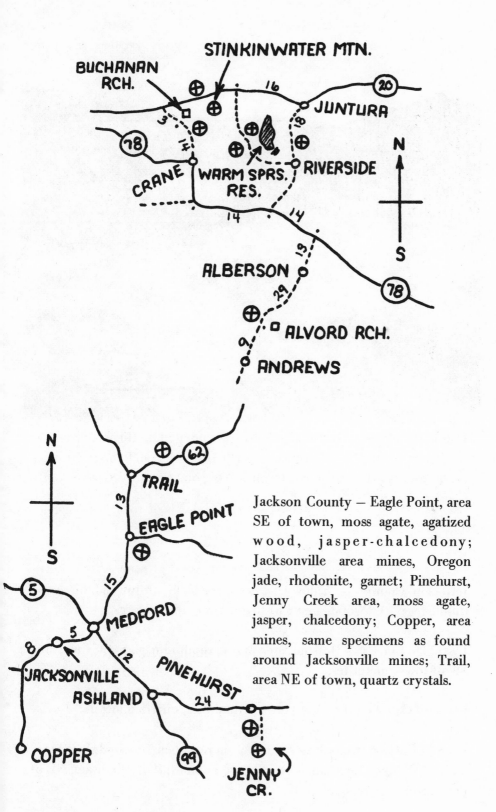

Jackson County — Eagle Point, area
SE of town, moss agate, agatized
wood, jasper-chalcedony;
Jacksonville area mines, Oregon
jade, rhodonite, garnet; Pinehurst,
Jenny Creek area, moss agate,
jasper, chalcedony; Copper, area
mines, same specimens as found
around Jacksonville mines; Trail,
area NE of town, quartz crystals.

Unusual geologic formations may be seen along State Highway 31, near the south end of Summer Lake, north of Paisley and about 40 miles north of Lakeview, in Lake County, Oregon. *Courtesy Oregon State Highway Photo 7506.*

Jefferson County — Willowdale, agate beds, S of town, blue agate, thundereggs; Ashwood, area S of town, agate, thundereggs.

Josephine County — Holland, area mines, rhodonite, garnet; Selma, area SW of town, metorites.

Klamath County — Bly, Quartz Mountain Pass, quartz minerals.

Lake County — Drews Reservoir, SW shores, opalized wood; Lakeview, area SE of town, at Crane Creek, agate nodules; Hart Mountain, NE of Plush, jasper-agate, opal.

Lane County — Lowell, area NW of town, agate, jasper.

Lincoln County — Agate Beach, S to Yachats, beach areas, petrified wood jasper, garnet.

An avid "rockhound" hammers at a boulder in the Ochoco Mountains east of Prineville, Oregon. Searching for, cutting, and polishing of agates, opals, quartz, garnet, and other precious stones is a popular hobby in Oregon. Mountainous and coastal areas of the state abound with a variety of such stones. This scene is at the Sheep Creek Agate Beds, maintained for free public use by the Prineville Chamber of Commerce. There are many other free and commercial rock hunting grounds in the area. *Courtesy Oregon State Highway Department Photo 7255.*

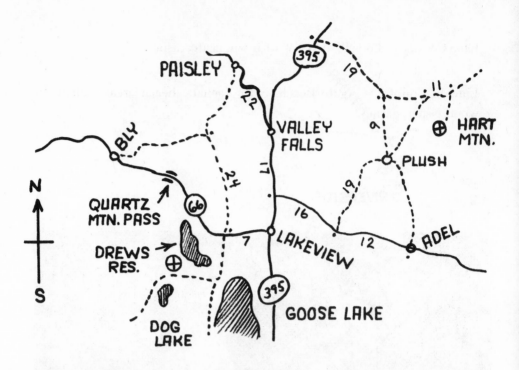

Geology's own record of the ages is contained in the fossil beds area of Grant County in eastern Oregon. Scientists say the formations date back for millions of years, based on the discovery of fossils of animals that became extinct before the dawn of history. *Courtesy Oregon State Highway Photo 183.*

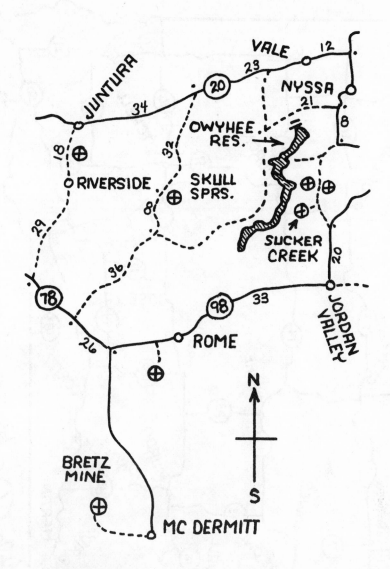

Malheur County — Bretz mine, NE of McDermitt, chalcedony, agate;
Rome, area SW of town, agate, petrified wood; Skull Springs, SW of
Vale, quartz crystals, jasper, thundereggs; Owyhee Reservoir, area E of
reservoir, opal, agate-wood; Sucker Creek area, agate thundereggs.

Sherman County — Wasco, area NW of town, picture jasper.

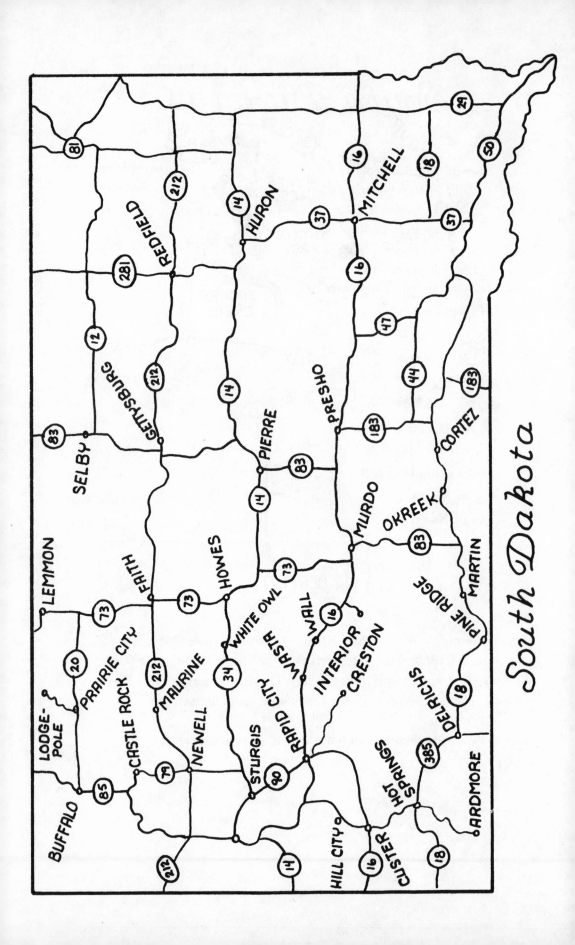

South Dakota

The state of South Dakota lies largely in the Great Plains except for the Black Hills in the southeast where the highest point of the Rockies rises to 7,240 feet. The eastern section of the state is rolling plain which gradually becomes hilly in the Coteau des Pairies. The area west of the Missouri is a dissected plateau, which includes the badlands area between the Cheyenne and White Rivers. The semiarid northwest regions of the state contain an area of gumbo soil and pierre clay.

NOTES OF INTEREST

Cassiterite is commonly found in the Black Hills of South Dakota.

Summer is considered the best collecting season in South Dakota.

In South Dakota's badlands crystals of barite have been found growing inside the hollow bones of fossil animals.

Bentonite is found near Belle Fourche, South Dakota.

The ranking gold mine of the western hemisphere is the Homestake Mine at Lead, South Dakota.

State geological surveys are available from the Bureau of Mines, Rapid City Experiment Station, School of Mines Campus, Rapid City.

The Black Hills of South Dakota are rich with rare fossils and quartz minerals.

Butte County — Arpan, NE shore area of Belle Fourche Reservoir, golden barite.

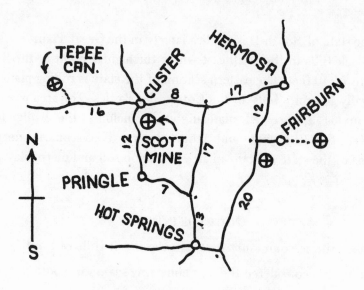

Custer County — Tepee Canyon, W of Custer, agate nodules; Scott Mine, SE of Custer, rose quartz; Fairburn, area E of town, jasper, agate, chalcedony, carnelian; area SW of Fairburn, agate.

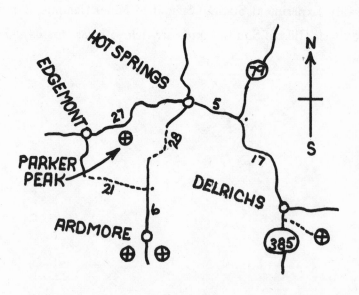

Fall River County — Ardmore, areas SW and SE of town, agate wood; Parker Peak, E of Edgemont, beryl crystals, petrified wood; Delrichs, area SE of town, agate, chalcedony, jasper-wood.

Mineral Occurrences

In South Dakota

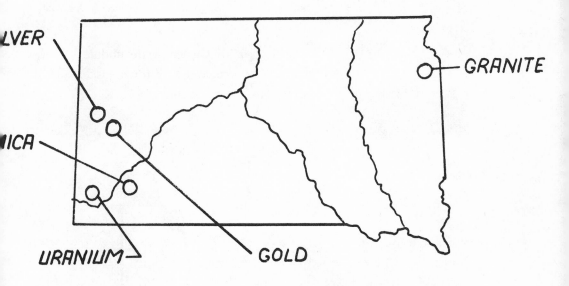

Jackson County — Interior, areas S and SW of town, chalcedony, carnelian, agate, jasper.

Lawrence County — Whitewood, area N of town, amethyst geodes.

Meade County — Sturgis, area SE of town, selenite crystals.

Standby Mine, Rockford, South Dakota.

Pennington County — Hill City, areas N and NE of city, rose quartz, tourmaline; Imlay, areas W and E of town, chalcedony, geodes; Scenic, area W of town, agate; Wasta, areas NW and N of town, golden barite, dog-tooth spar.

Perkins County — Lodgepole, area NW of town, petrified wood; Lemmon, large area W of town, petrified wood.

NOTES OF INTEREST

Wolframite had been found to some extent in the Black Hills of South Dakota.

The South Dakota School of Mines and Technology in Rapid City has a notable mineral exhibit.

Hunting rocks in South Dakota. *Courtesy Travel Division, South Dakota Department of Highways.*

South Dakota rock crystals. *Courtesy Travel Division, South Dakota Department of Highways.*

Amber-colored barite crystals are often found teamed with snaillike fossils in the concretion formations of western South Dakota. *Courtesy Travel Division, South Dakota Department of Highways.*

Pierre, South Dakota. You never know what the next rock will be, and this one happened to be a real prize. Fairburn agates like this one are found in several places in South Dakota. *Courtesy Travel Division, South Dakota Department of Highways.*

Homestake Mining Company surface plant at Lead in Lawrence County, South Dakota. Major plant units are: Yates shaft (upper left); mechanical department's machine shops, foundry, pattern shop, and warehouse (below and to right of Yates shaft headframe); south mill (left center); east and west cyanide plants (flat-roofed structure at lower left and building at lower right); refinery (immediately above west cyanide plant and to right of south mill). *Courtesy Homestake Mining Co.*

Standby Mine, Rockford, South Dakota.

Rock collecting in South Dakota. *Courtesy Travel Division, South Dakota Department of Highways.*

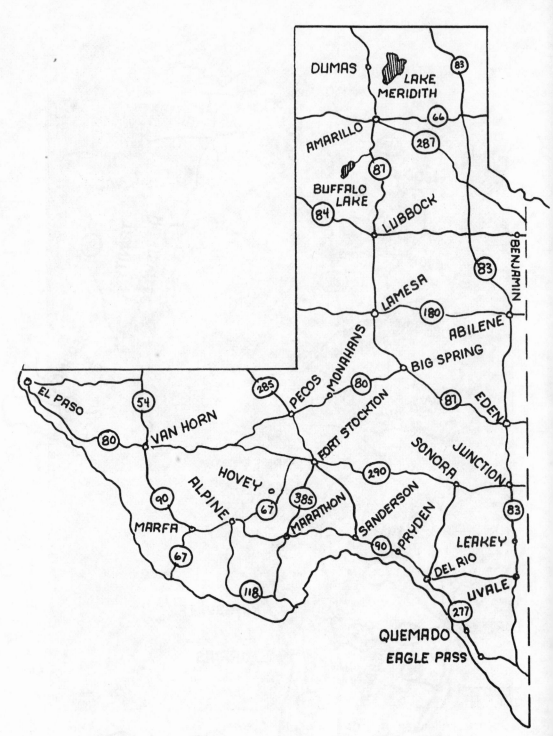

West Texas

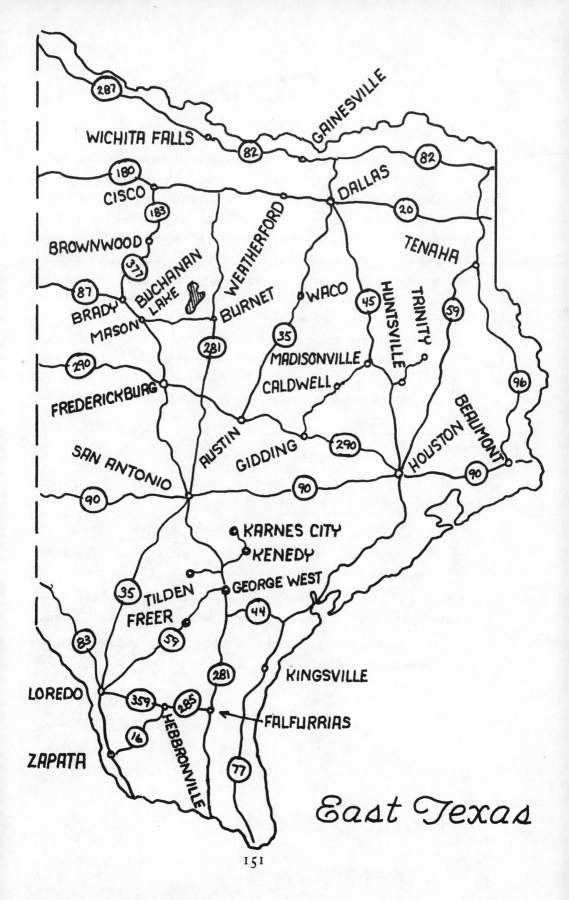

East Texas

Texas

Along the western border of Texas is the black prairie belt, consisting of rich black limestone soil, while eastern Texas consists of low, sandy hills and rolling plains, extensively timbered. West of Dallas there is a transition zone between the coastal plain and the interior lowlands, comprising the Grand Prairie and is flanked by the Eastern and Western Cross Timbers. Extending south, from central Oklahoma, are the interior lowlands, rolling prairie land where there is occasional resistant limestone and sandstone escarpments. In the Panhandle region and west is the High Plains section of the Great Plains, a semiarid, treeless, almost level expanse, bordered on the east by the irregular Cap Rock escarpment. A southeastern extension of the High Plains is the Edwards Plateau, another resistant limestone formation. This plateau is cut off from its western section, the Stockton Plateau, by the deep Pecos River canyon.

NOTES OF INTEREST

The gravel beds of the Delta, from Laredo to Falcon Dam, yield enormous quantities of agate of every description.

Large crystals of celestite have been found at Lampasas, Texas.

The best collecting seasons for most areas in Texas are fall, winter, and spring as the summers are extremely hot.

State geological surveys are available from the Bureau of Economic Geology, University of Texas, University Station, Box 8022, in Austin.

Mason County, Texas, is considered to be an outstanding collecting locality for topaz.

Zapada County — Zapada, Rio Grande River areas SE and W of town, petrified wood, agate.

Webb County — Singing Hill Ranch, NW of Laredo, agate; Pico Ranch, SW of Singing Hill Ranch, agate.

Maverick County — Eagle Pass, Helms Ranch, NW of town, moss and plum agate; Quemado, Villarreal Ranch, NE of town, agate-chert.

Real County — Leaky, Horsecollar Park area, NE of town, calcite crystals, geodes.

Mineral Occurrences In Texas

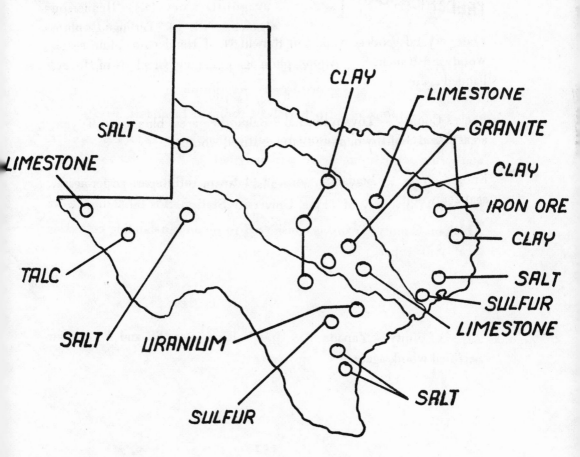

Mason County — Mason, mining claim W of town, topaz; Katemcy Creek, 10 mi. N of Mason, topaz and smoky quartz.

Llano County — Llano, Enchanted Rock, SW of town, green epiodite, quartz crystals; area N of Llano, along both sides of hwy, pink feldspar, llanonite.

Terre County — Dryden, Kothman Ranch, SE of town, agate-wood.

Brewester County — Marathron, Love and Stillwell Ranches, SE of town, agate; Terlinqua, Needle Peak, SW of town, sagenite agate, aragonite crystals; Henderson Ranch, NE of Terlingua, plum agate; crystal geodes; Anderson Ranch, N of Henderson, plum agate; Woodward Ranch, S of Alpine, plum agate; Paisano Creek, N of Hovey, banded agate.

Reeves County — Toyahvale, Lake Balmorhea area, blue agate; Toyah, areas N and S of town, plum agate, agatized wood.

Knox County — Benjamin, area N of town, satin spar, jasper-agate, agatized wood.

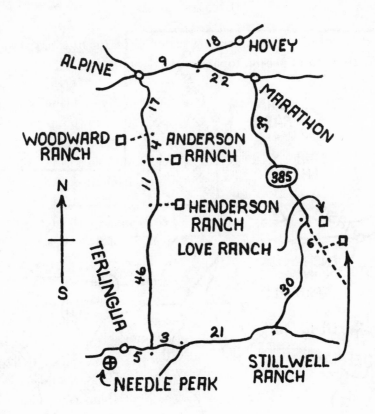

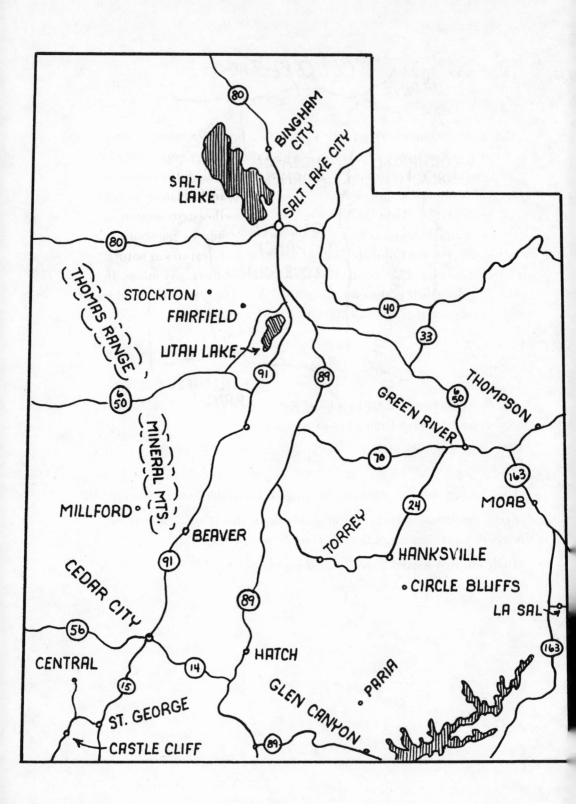

Utah

Utah

East-west mountain ranges, desert basins, broad tablelands, deep canyons, and river valleys comprise Utah's varied topography. A series of broken ranges and plateaus extend north-south through the center of the state, forming the boundary between the Colorado Plateau in the east and the Great Basin on the west. In the northeastern section of Utah the Uinta Mountains rise to 13,498 feet, while to the south in eastern Utah, lies the Colorado Plateau, offering such features as natural bridges, multicolored sandstone cliffs, and isolate buttes and mesas. In western Utah, which is the eastern part of the Great Basin, the land is drab, with its desert plains, salt flats, and block mountains.

NOTES OF INTEREST

An enormous concentration of uranium is located in the Chinle and Morrison sandstones in Utah's Four Corners region.

The state's mining districts are in the vicinity of Bingham Canyon which has a large open-pit copper mine.

Utah's most valuable minerals are copper, lead, silver, zinc, and gold.

State geological surveys are available from the Geological Survey, Mineral Deposits Branch, 222 S.W. Temple, in Salt Lake City.

Utah was first named Deseret, meaning "honey bee."

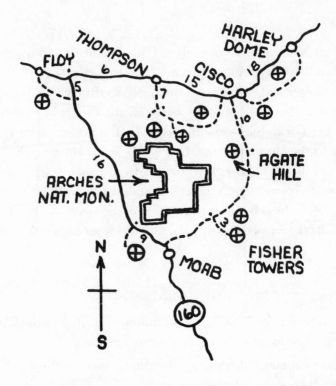

Grand County — Moab, area NW of town, chalcedony, petrified wood; Fisher Towers, N of Castleton, petrified wood, quartz; Agate Hill, S of Cisco, agate, jasper; Cisco, areas E and SE of town, agatized wood; Thompson, area S and SE of town, toward Arches National Monuments, agate; Floy, areas S and SE of town, agate.

Juab County — Jericho, areas NW and W of town, agate, jasper; Callao, Bennion Ranch, SE of town, agate; Topaz Mountain, area SE of Bennion Ranch, chalcedony, geodes, garnet, topaz; Dugway Mountains, NE of Callao, agate, crystal geodes.

Mineral Occurrences In Utah

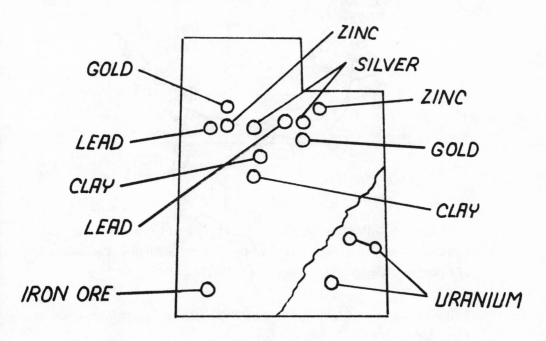

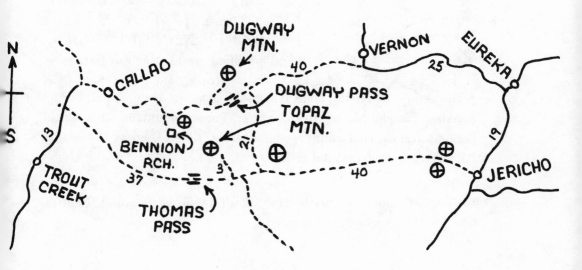

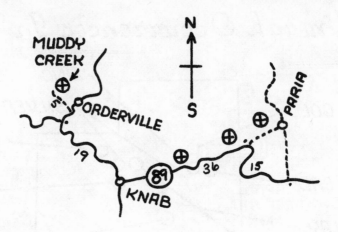

Kane County — Porderville, Muddy Creek, NW of town, calcite crystals, septarian nodules; Kanab, area E of town, petrified wood; Paria, area W of town, petrified wood.

Millard County — Black Rock area NE of town, snowflake obsidian; Clear Lake, area S of town, sunstones.

San Juan County — La Sal, area NW of town, jasper wood; Natural Bridges National Monument, petrified wood; Mexican Hat, areas E of town, along the San Juan River, agate.

Washington County — Leeds, area SE of town, agate wood; Central, area NW of town, agate, geodes.

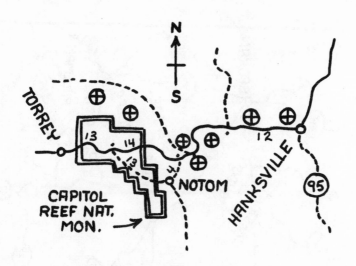

Wayne County — Hanksville, area N of town, petrified wood; Fremont River area, W of Hanksville, agate, jasper, petrified wood; Notom, area NE of the town, chert, calcite crystals; Capitol Reef National Monument, area NE of the monument, agate nodules, selenite crystals.

Weber County — area mines, malachite, azurite.

NOTES OF INTEREST

Deposits in Utah, at Bingham Canyon, produce substantial amounts of enargite.

Peach-colored crystals of topaz can be found in the Thomas Range of western Utah.

Spring and fall are the best collecting seasons in southern Utah, while late spring, summer, and fall are considered best for the northern regions of the state.

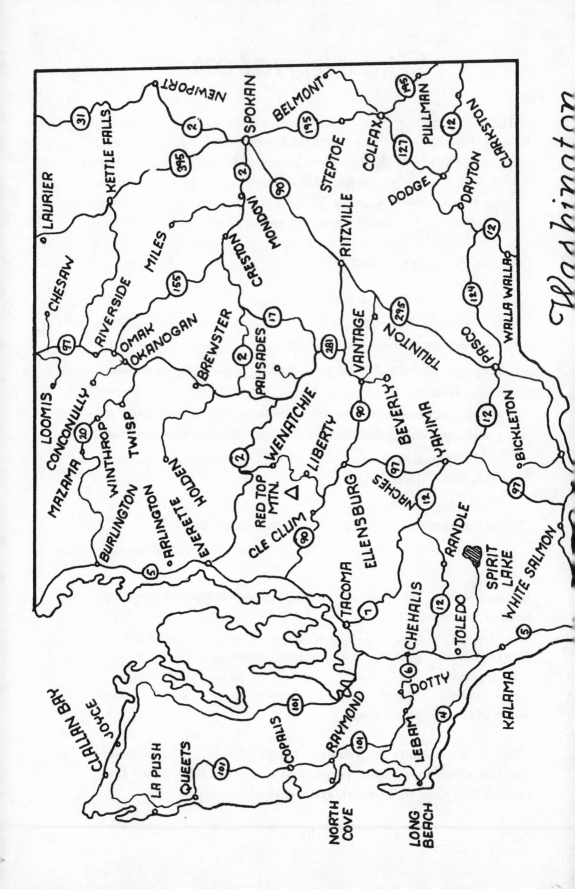

Washington

The topography of Washington is remarkably varied and bold. The Cascade Mountains, running north and south, divide the state into a green, wet western region and an arid, canyon-cut eastern region. Puget Sound, with more than 300 forested islands, bisects the western portion, creating the Olympic Peninsular. To the south Grays Harbor and Wellapa Harbor, reaching far inland from the sea, are surrounded by heavily timbered hills. Eastern Washington forms the major portion of the Columbia Plateau, a vast, arid, almost level region of Miocene basalts.

NOTES OF INTEREST

The Columbia River beaches yield quantities of gemstones and prehistoric Indian artifacts such as obsidian arrowheads.

Late spring to early fall are the best seasons for collecting in most areas of Washington.

Throughout western Washington, the road cuts stream beds, and fields produce prized agatized clams, oysters, and other Miocene fossils.

The Washington State Museum, University of Washington in Seattle offers a notable mineral exhibit.

State geological surveys are available from the Geological Survey, Mineral Deposits Branch, South 157 Howard St., Spokane.

Callam County — La Push area beaches, jasper, chalcedony; Joyce area beacher, jasper pebbles.

Cowlitz County — Kalama, area E of town, moss agate, carnelian.

Grant County — Beverly, area SE of town, opalized wood; Taunton, area S of town, agatized wood.

Gray Harbor County — Taholah, Pacific Beach and Copalis area beaches, chalcedony, agate jasper, and quartz pebbles; North Cove, beaches N of Cove to Westport, agate, chalcedony, jasper.

Jefferson County — Queets area beaches, to Taholah, jasper-agate, orbicular jasper.

Kittitas County — Cle Elum Lake, area S of lake, agate geodes. Peoh Lookout, S of Cle Elum, agate geodes; Liberty, Red Top Mountain NW of Liberty, gem agate; area S of mountain, jasper, moss agate, geodes; Table Mountain, NE of Liberty, blue agate; W. Beverly, S of town, petrified wood.

Mineral Occurrences
In Washington

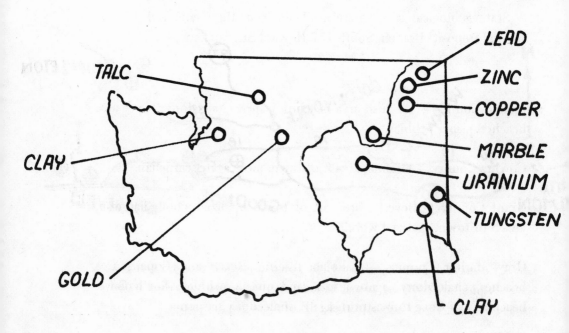

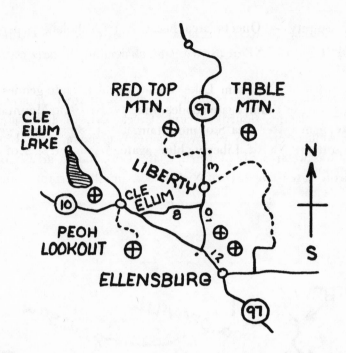

Klickitat County — Warwick, area W of town, jasper wood, carnelian; Goodnoe, area N of town, chert; Bickleton, area SW of town, along hwy, jasper, carnelian; Roosevelt, areas NW and NE of town, opalized wood.

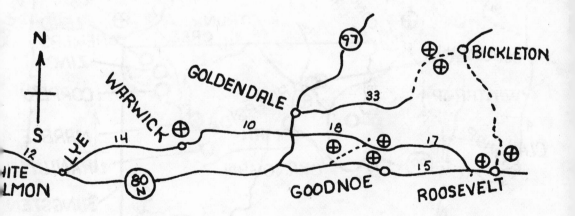

Lewis County — Toledo, area SE of town, petrified wood, carnelian, agate; Alpha, area NW of town, bloodstone, carnelian, petrified wood, geodes; Randle, East Canyon Creek, SE of town, agate, jasper.

Lincoln County — Reardan, area N of town, near Spokane River, pastelite; Miles, area NE of town, agate, chalcedony, quartz crystals.

Okanogan County — Riverside, area S and W of town, quartz, garnet, tourmaline, clacite, flourite; Trunk Creek, NE of Riverside, mica, epidote; Aeneas, area NW of town, agate; Wauconda, area NE of town, garnet; Chesaw, area SE of town, agate; Conconully, area N of town, jasper; Iron Gate Camp, SW of Nighthawk, amethyst, smoky quartz.

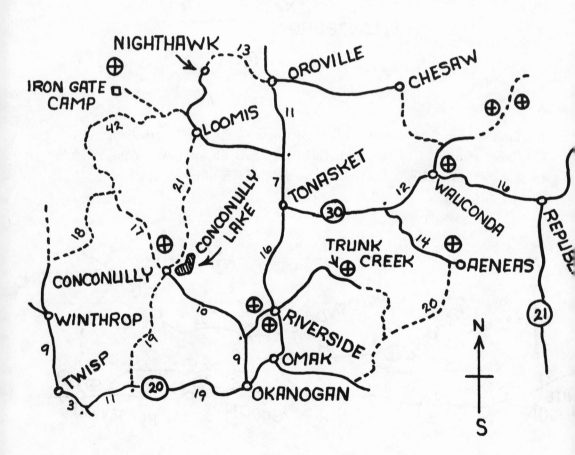

Pend Oreille County — Metaline Falls, Josephine Mine, smithsonite; Calispell Peak, SW of Tiger, beryl, muscovite; Sacheen Lake, W of Newport, garnet.

Yakema County — Wiley, Ahtanur Ridge, jasper; Moxes, area N of town, at Yakima Ridge, petrified wood.

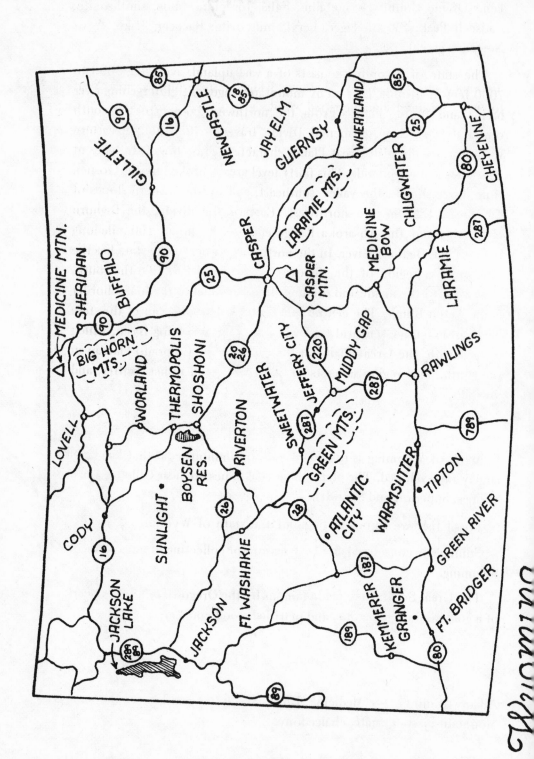

Wyoming

The state of Wyoming consists of a vast upland, averaging 5000 to 7000 feet in altitude, crossed by mountain ranges with intervening river basins and rolling plains. From the northwest corner to the south central border the Continental Divide traverses the state. In eastern Wyoming are the high Great Plains, characterized by broad stretches of grasslands and woodlands. This fairly level area is broken by the Goshen Hole depression in the valley southeast, and by the western slopes of the Black Hills in the northwest. East of the divide, the Bighorn Mountains and the Absaroka Range enclose the broad 100-mile-long basin of the Bighorn River. In the northwest corner of the state lies the lofty plateau region of the Yellowstone National Park. To the south, lying west of the divide and running almost parallel to the Idaho border is the Teton Range. The central and southwestern portions of the state lie in an extensive semiarid tract known as the Wyoming Basin, which merges with the Great Plains in the east and the Colorado Plateau in the south. This region consists of sagebrush, sand dunes, buttes, and low hills.

NOTES OF INTEREST

Much of Wyoming is underlaid with bituminous coal that has been largely undeveloped. The states main coal mines are near Gillette, Rock Springs, Superior, and Sheridan.

Fossil fish are found throughout the state of Wyoming.

Summer is considered the best season for collecting in most areas of Wyoming.

The Great South Pass, made famous by the Oregon Trail, is the heart of a gold mining, gemstone, and petrified wood country.

Albany County — Medicine Box, NE on W slopes of Laramie Mountains; jasper, agate, chalcedony.

Big Horn County — Cloverly, area S of town, petrified wood; Hyatt-ville, Spanish Point, NE of town, chalcedony, moss agate.

Carbon County — Riverside, area mines, quartz crystals, garnet, amethyst, azurite, chalcocite; Saratoga, areas N of town, opalized wood, chalcedony, agate; Seminoe Reservoir, area NW of reservoir, jade.

Freemont County — Crowheart, Wind River area gravels, petrified wood, agate, jasper; Riverton, area SW and S of town, moss agate; Sweetwater Station, area SW of station, moss agate; Jeffery City, area NE of city, wyoming jade; areas S and SW of Jeffery City, jade; Crooks Gap, S of Jeffery City, agate, jade.

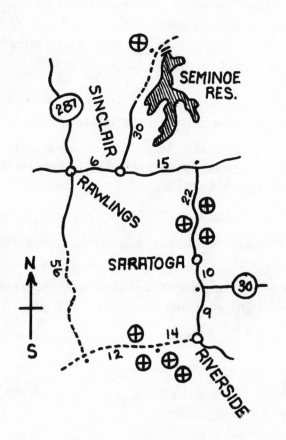

Mineral Occurrences In Wyoming

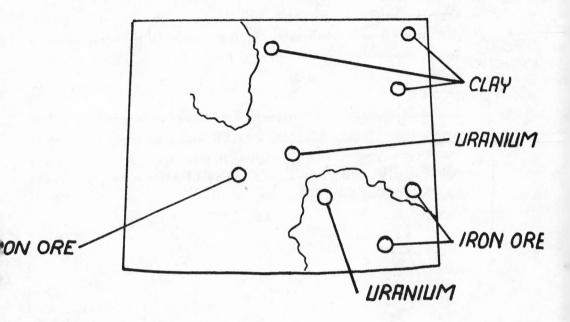

Natrona County — Casper, area quarrys, amozonite, agate.

Platt County — Glendo Reservoir, Eloxite Claim, NW of reservoir, common opal, agate-jasper; Hartsville, agate mine, S of town, onyx mine NE of Hartsville.

Sweetwater County — Greenriver, Whiskey Basin, NW of town, agate wood; Farson, Oregon Butte and Eden Valley, NE of Farson, agate-chalcedony, petrified wood; Tipton, area S and SE of town, petrified wood, turritella agate.

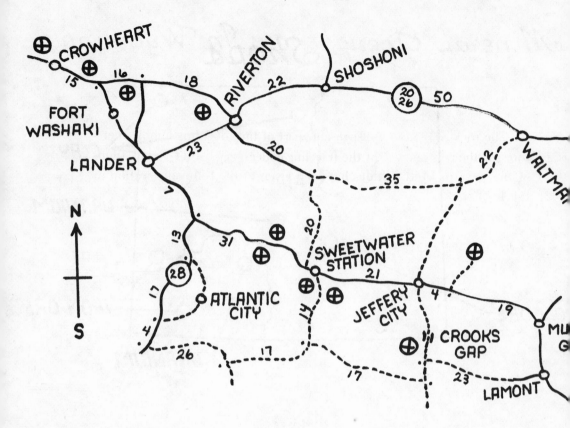

Rock Shops

The rock shop is an indispensable part of the collecting hobby, therefore for the convenience of the traveling collector, a listing of rock shops in the western United States has been given in the following section of this book.

Arizona Rock Shops

APACHE JUNCTION
Fredrich's

GLOBE
Copper City Rock Shop
566 Ash Street

PHOENIX
Farmers Gem & Rock Shop
10037 Cave Creek Road

Arrow Gems & Rock Shop
9827 Cave Creek Road

M & R Rock Shop
2225 N. 23rd St.

QUARTZITE
Paul's Lapidaries

SEDONA
The Red Rock Shop

Ramsey's Rocks & Minerals
227 N. Hwy. 89 A

ST. JOHNS
St. John's Rock Shop

TOMBSTONE
Tombstone Rock Shop
506 Allen Street

TUCSON
Rocks & Minerals
125 E. Grant

Continental Minerals
4737 E. Adams Street

Pardo's Rocks & Minerals
6756 Nogales Hwy.

Evergreen Agate Shop
4450 E. Bermuda

WICKENBURG
Dunkin's Rock Pile
Prescott Hwy N.

Webbs Rock & Gem Shop

YARNELL
Navajo Trail Gem Shop

YUMA
Duncan's Minerals
2424 Kathleen Ave.

Navajo Center Art & Gem Shop
6 mi. E of Yuma

California Rock Shops

AUBURN
Re-Peat Rock Shop
105 Midway

ACTON
Tepee Rock Shop
9750 Soledad Canyon Rd.

APPLE VALLEY
Wash-Dak-Cal Gemshop
22745 Romona

ATASCADERO
Western Gems
9465 Morro Rd.

ARROYO GRANDE
Dad's Rock Shop
112 Cherry Ave.

AZUSA
Murray's Lapidary Supply
625 N. Azusa Ave.

BLYTHE
J & D Gem Shop
Hwy 95

COMPTON
Compton Rock Shop
1405 S. Long Beach Bl.

DOWNEY
Shaw's Rock Hut
8822 E. Imperial Hwy.

FORTUNA
Humboldt Rock Shop
12 N. Main St.

HAWTHORN
Fran's Gem Shop
3735 Rosecrans

LA PUENTE
Kem's
14039 Don Julian Rd.

The Rock Shop
15830 E. Main

PALMDALE
Hicks Hobbies Lapidary
2534½ E. Ave. Q

PORTERVILLE
Gene's Rock Shop
1155 N. Grand

POWAY
Schneider's Rocks & Minerals
13021 Poway Rd.

RIDGECREST
Indian Well Lapidary Shop
1234 W. Ridgecrest Bl.

RIVERSIDE
Shamrock Rock Shop
593 West La Cadena Dr.

ROWLAND HEIGHTS
CALIFORNIA Gemcrafts
1403 Almena

SACRAMENTO
Batten's Gem Center
2631 El Paso

SAN BERNARDINO
Wergin's Rock
18572 Cajon Blvd.

SAN DIEGO
The Geode Shop
6080 University Ave.

SAN CLEMENTE
Gemart
3607 S. El Camino Real

SAN LEANDRO
Aladin's Rock Shop
1801 E. 14th St.

SAN JUAN BAUTISTA
Poppy Jasper Rock Shop
25 Muckelemi

STOCKTON
Western Gemstones
3245 Waterloo Rd.

SAN MARCOS
The Rock Farm
1145 Law St.

TEHACHAPI
Gil's of Tehachapi
114 West G St.

TWENTYNINE PALMS
B&B Rock Shop
6451 Adobe Rd.

TURLOCK
Turlock Lapidary
248 W. Main

VACAVILLE
TNT Rock Shop
Brown's Valley Rd.

VAN NUYS
San Fernando Valley Gem co.
5905 Kester Ave.

VISALIA
Visalia Rock & Gem Mart
29710 Road 80

Colorado Rock Shops

BAYFIELD
Shipleys Mineral House

BOULDER
The Gem Trail
3304 Pine Street

COLORADO SPRINGS
Ackleys Rocks
3910 N. Nevada

Baders Minerals
836 N. Institute

DENVER
Agate Shop
329 S. Pearl

ESTES PARK
Dicks Rock Museum
Moraine Park Rd.

FORT COLLINS
Jade - N - Gems
518 Remington

GRAND JUNCTION
Mt. Garfield Lapidary

GUNNISON
Gunnison Rockery
118 N. Wisconsin

OURAY
Columbine Mineral Shop

MANITOU SPRINGS
Vista Verde Rock Shop

PONCHA SPRINGS
Bob's Rock & Gift Shop

PUEBLO
Mesa Gems & Minerals

Idaho Rock Shops

BURLEY
Arco Rock Shop
Hwy 25

BOISE
Stewart's Gem Shop
2620 Idaho Street

CALDWELL
Gene's Rocks & Gems
1824 S. Kimball Ave.

LEWISTOWN
Butler's Rock & Craft Shop
910 Prospect

MC CALL
Ron's Gem Shop
Warren Wagon Road

NAMPA
Beail Rock Shop
1810 Sunny Ridge Road.

NORTH FORK
Maydole Rock Shop

OROFINO
Gem State Gems
Hwy 12

TWIN FALLS
Moon's
301 West Main

VICTOR
Gillette's Gems & Gift Shop

Kansas Rock Shops

ALMA
Shumake Rock Shop

ABILENE
Mc Kee Gardens Gem-Craft
1524 W. 1st.

GALENA
Boodle Lanes Rock & Mineral
Shop

ISABEL
K - 42 Rock Shopp

KINGMAN
Hobby Rocks
1040 Hwy 54 E.

MARYSVILLE
Jim & Angie's Rock & Hobby
Shop, 1205 Laramie

WELLINGTON
Dasher's Rock Shop
221 W. Maple

WICHITA
Ebersole Lapidary Supply
11417 W. Hwy 54

Montana Rock Shops

BUTTE
Trevillion-Johnson Gems
2400 S. Montana Ave.

Koop's Gems
3348 Harrison Ave.

CRANE
Mary Ann's Agate & Gift Shop

ENNIS
Ennis Rock House

GREAT FALLS
Montana Gem & Lapidary
U.S. 87 N.

Al's Rock Shop
No. 11 Holiday Village

HAMILTON
Hamilton Montana Lapidary
515 Marcus

KALISPELL
Edward's Gem Shop
512 2 Ave. East

LAURIN
Sam's Agate House

LEWISTOWN
Yogo Gem & Rock Shop

MILES CITY
Fry's Gem Shop
1207 Woodbury Street

MISSOULA
Bitterroot Lapidary

Randy's Rock Shop
700 S. 3rd.

POOLSON
Stuver's Rock Shop
Hwy. 93 & Bayview Dr.

SIDNEY
Frandsen's Agate & Gift Shop
606 S. Central Ave.

SILVER STAR
Brant's Rock Shop

TERRY
Rahl's Agate Store

Nevada Rock Shops

BATTLE MOUNTAIN
Todd's Rock Shop

DENIO
Eaton's Wagon Wheel Motel
& Rock Shop

FALLON
Highway 50-95 Rock Shop
4261 Reno Hwy

IMLAY
Rocks & Gems

LAS VEGAS
Gem Dandy
1701 East Charleston

Continental Gems & Minerals
2007 E. Charleston Blvd.

RENO
C & H Lapidary
1585 S. Wells

Covered Wagon Gems

New Mexico Rock Shops

ALAMOGORDO
Desert Lapidary
3005 N. White Sands Blvd.

ALBUQUERQUE
Sun Country Minerals

Mc Crory's Rock Shop
400 Santa Fe N.W.

Kohl's Rocks & Minerals
932 Eubank

Mc Kee Rock Garden & Museum
331 Nara Visa Rd.

DEMING
Krol's Rock City
State 26

Deming Rock Shop

FARMINGTON
The Colorado Gem Company
2328 East 14th Street

GILA
Bear Creek Rock Shop

LORDSBURG
Border Rock Shop
980 E. Railroad

MESILLA
Pebble Pups Shop
13 Hwy 28

Santa Fe Gem & Mineral Shop
3151 Cerrillos Road

WHITE CITY
Ray's Rock Shop

Oklahoma Rock Shops

BARTLESVILLE
Windle's Rock Shop
Hwy 75

OKLAHOMA CITY
Arrowhead Lapidary
330 S.W. 28th

EL RENO
Muston Rock Shop
300 N. Beckford

TULSA
Treasure Shack
8500 E. 11th Street

Oregon Rock Shops

AGATE BEACH
Wolf's Rock Shop

ARCH CAPE
Far West Rocks

ASHLAND
Dab's Rocks
2785 E. Main St.

BEND
McConville Rock Shop

BURNS
Highland Rocks & Gift Shop
1316 Hines Blvd.

CAVE JUNCTION
Dryden's Rock Shop

EUGENE
Ray's Rock Shop
4952 River Rd.

Oregon Lapidary Supply
605 W. 4th Ave.

GRANT'S PASS
John Bastian Rock Shop
5335 Upper River Rd.

LAKEVIEW
Daves Rock Shop
764 S. F. Street

MEDFORD
The Sanvigs
902 N. Riverside

PRINEVILLE
Elkins Gem Stones
833 S. Main St.

ROSEBURG
Le Blue's Rock Shop
310 SE Jackson St.

WYETH
Wind Mt. Rock Shop

YACHATS
Earl's Agate Shop

COOS BAY
Wayside Rock Shop

South Dakota Rock Shops

BOX ELDER
P & K Lapidary

CUSTER
Scott's Mineral Box
1020 Custer Ave.

DEADWOOD
76 Rock Shop

Miller's Rock Shop

KEYSTONE
Gems & Minerals

RAPID CITY
Town N Country Rock Shop
Mt. Rushmore Road

Fairburn Agate Shop
2114 Mt. Rushmore Road

Scheirbeck's Rock & Fossil
Shop

Texas Rock Shops

ALPINE
Twin Peaks Motel & Rock Shop
Hwy 90 W.

AMARILLO
Goodnow Gems
3608 Sunlite

BEAUMONT
W Rock Shop
4590 Magnolia

DALLAS
Bishop's House of Gems

Lapidary Workshop
3418 Greenville Ave.

EL PASO
Gem Center U.S.A.
4100 Alameda

Aztec Calendar Rock Shop
4422 Clifton

Davis Gem Craft
2801 Montana Ave.

FORT WORTH
Melbourn Gem Company

FLORESVILLE
L & M Rocke & Cermic Shop
Hwy 181 N.

GIDDING
House Of Gems
Hwy 290 E.

HOUSTON
Larkin Lapidary
4023 Westheimer

Fred & Sue's
8338 Park Pl. Blvd.

King's Gem Center
5409 Nordling

J & L Lapidary
8335 Findlay Street

LEAGUE CITY
Bauer Rock & Seashell Shop
204 N. Hwy.

LEWISVILLE
P & P Rockcraft

LUBBOCK
Youngs Rocks & Gems

SAN ANTONIO
Fessman's Rock Shop
339 Donaldson

WICHITA FALLS
Smith Rocks & Minerals
1525 Beverly Drive

ZAPATA
Zapata Rock Shop

Utah Rock Shops

ELSINORE

Johnny's Rock Shop
Hwy 87

MOAB

Moab Rock Shop
137 N. Main

ORDERVILLE

Joe's Rock Shop
Hwy 89 N.

PROVO

Stan's Shop
123 West 500 North

SALT LAKE CITY

Ken Stewart's Gems
220 W. Third Street

Dowse's Rock Shop
754 North 2nd W.

ST. GEORGE

Eutsler's Rock Shop
W. Hwy 91

Washington Rock Shops

CHEHALIS
Adames Gem & Gift Shop
Market St.

KELSO
Lance's Lapidary
700 S Pacific

KIRKLAND
Chuck Havlik's Rock Shop
10619 NE -116

MILLWOOD
Valley Rock Shop
E 9319 Trent

NORTH BEND
D B Gems

OKANOGAN
Fran & Ollie's Rock Shop
123 1 st. Ave.

PASCO
Highland Gems & Minerals
1147 Columbia Dr.

PUYALLUP
L & H Rock Shop
2220 W. Pioneer

POST FALLS
Silver Capital Gems

RENTON
Polson's Rock Shop
4443 S 166th St.

TACOMA
Lakewood Lapidary
1230 Lake City Blv.

Clark's Agate
219 S 50th

VICTORIA
Rockhound Shop
850 Tolmie

WESTPORT
Jim's Agate Shop

YAKIMA
Liberty Rock Shop

Wyoming Rock Shops

CASPER
The Jade Place
1765 W. 15th

CHEYENNE
Glenella Stone Hut
104 Third Ave.

GREEN RIVER
Wyoming Rock Shop

JACKSON
Teton Gemstone Shop

LARAMIE
Clark's Gem Material
1404½ Park Ave.

LANDER
Norman's Rock Shop
240 N. 8th Street

Tornado Rock Shop
Snavely Lane

NEW CASTLE
Mel's Rock Shop
531 W. Warren

POWELL
The Brutweisers
510 N. Division

RAWLINGS
Walker's Rock Pile
1122 11th Street

Round's Jade
820 North Seventh

TORRINGTON
Torrington Rock Shop
Rt. 1

WORLAND
Bessons Rock Shop
621 Howell Ave.

Campgrounds
And Motels

For the convenience of the traveling collector, there are campground and motel accommodations for each state listed in the following section of this book. Only those accommodations within easy reach of collecting sites have been listed and although every attempt has been made to establish the accuracy of this listing, the reader must remember that this is an age of change and what was a motel today may be a shopping center tomorrow.

Arizona Campgrounds

BENSON
Casey's Trailer Court, E on 1-10 and State 86.

BULLHEAD CITY
River Green Resort, 3 mi. S on the Colorado River.

CAVE CREEK
Cave Creek, 20 mi. N on Forest Road 24.

CIRCLE CITY
Circle City KOA, in town.

CLINTS WELL
Blue Ridge, 10 mi. NE on Recreation Route 10.

DOLAN SPRINGS
Dolan Springs Trailer Park, in town.

DOUGLAS
Bathtub, 32 mi. NE on U.S. 80, then 16 mi. W on county road, Forest road 74.

FLAGSTAFF
Ashurst, 19 mi. SE on Forest Hwy. 3 and Mormon Lake-Long Valley road.

GLOBE
Jones Water, 17 mi. NE on U.S. 60.

KINGMAN
Cerbat Mountain, 27 mi. NW via Chloride on county road.

MIAMI
Warnica Springs, 3 mi. S on Forest Road 580.

NOGALES
Pena Blanca, 6 mi. NW on U.S. 89, then 11 mi. SW on Ruby Rd.

ORGAN PIPE CACTUS N.M.
Headquarters Campground, 35 mi. S of Ajo on State 85.

PAGE
Chapman's Trailer Park, 1¼ mi. off U.S. 89.

PARKER
Branson's Resort, 8 mi. N on State 95, on the Colo. River.

PAYSON
Christopher Creek, 21 mi. NE on Forest Hwy 11.

QUARTZITE
Crystal Hills, 17 mi. S on State 95, then 4 mi. E via county road.

SAINT DAVID
Ashcroft's Golden Bell Park, 1½ mi. S on U.S. 80.

SAINT JOHN'S
Lyman Lake State Park, 10 mi. S on U.S. 180 & 666, then 1½ mi. E on State 81.

SEDONA
Banjo Bill, 9 mi. N on U.S. 89-A, in Oak Creek Canyon.

SUPERIOR
Oak Flat, 5 mi. E on U.S. 60-70-89.

TUCSON
Bear Wallow, 39 mi. NE on Hitchcock Forest Hwy. 33.

WINSLOW
Rock Crossing Campground, 55 mi. S on State 63.

YUMA
Fisher's Landing, 24 mi. N via U.S. 95, then 10½ mi. W on Martinez Lake Rd.

California Campgrounds

AGUANGA
Dipping Springs, 7 mi. W on State 71.

ALPINE
Alpine Trailer Campe Rancho Resort, 2½ mi. E via U.S. 80, then 2½ mi. NE on Willows Rd.

ANGEL'S CAMP
Calaveras Big Tree State Park, 24 mi. E on State 4.

BAKERSFIELD
Dan's Trailer Lodge, 3 mi. SE on State 99.

BARSTOW
Calico Ghost Town Regional Park, 8 mi. NE off 1-15.

BIG BAR
Big Flat, 3½ mi. E on State 299.

BIG PINES
Apple Tree, 2 mi. NW on Pearblossom Rd.

BISHOP
Big Meadow, 22 mi. NW on U.S. 395, then 4½ mi. S of Tom's Place on Rock Creek Rd.

BLYTHE
Cibola Lake Club, 33 mi. SW via Cibola Toll Bridge & Cibola Lake Rd.

BORREGO SPRINGS
Anza-Borrego Desert State Park, Surrounding town off State 78.

CRESTVIEW
Big Springs, 1½ mi. S on U.S. 395, then 2 mi. E on Big Springs Rd.

EL CENTRO
El Centro Travel Trailer Lodge, 2 blks. E on U.S. 80 from the junction with State 86.

EMIGRANT GAP
Fuller Lake, 4 mi. N on forest road to Fuller Lake.

EUREKA
E-Z Boat Landing & Trailer Court, 3 mi. S on U.S. 101, then turn on King Ave. 1½ mi. past Shipreck Aquarium.

FRESNO
Millerton Lake, 29 mi. NE off 145.

GABERVILLE
French's Tourist Center, 8 mi. S on U.S. 101.

HAVILAH
Breckenridge, 2½ mi. S then 11 mi. W.

HAYFORK
Philpot, 8 mi. S on State 36.

HEMET
Roseland Mobile Homes, 4½ mi. W on State 74.

JULIAN
Cuyamaca Rancho state Park, 8 mi. S on State 79.

JUNCTION CITY
Junction City, 1¼ mi. NW on U.S. 299.

KLAMATH
Blackberry Patch Trailer Park, E off U.S. 101 at Terwer Valley turn-off to Klamath Glen .

LAKE HENSHAW
San Luis Rey, 3 mi. NW on State 76.

LAKE ISABELLA
Boulder Gulch, 4 mi. N on State 178, on Isabella Res.

LEE VINING
Big Bend, 5 mi. W on State 120, then 2 mi. W on Lee Vining Creek Rd.

LITTLE LAKE
Fish Camp, 7 mi. S on U.S. 395, then W on Ninemile Canyon Rd.

MOJAVE
Sierra Trails Lodge, 6½mi. N on State 14.

NAPA
Lake Berryessa Marina, in town at 5800 Knoxville Rd.

NEVADA CITY
Skillman Flat, 16 mi. E on State 20.

PALO VERDE
Walter's Camp, 12 mi. S on county S78, then 7 mi. E.

PASO ROBLES
Rest Haven Park, 2 mi. W then 24th St. to Adelaide Rd.

PINE VALLEY
Boulder Oaks, 9 mi. SE on U.S. 80.

PLACERVILLE
Finnon Lake, 8 mi. N off U.S. 50.

PORTERVILLE
Recreation Area No. 4, 5 mi. E on State 190.

RAMONA
Black Canyon, 11 mi. N on Ramona-Mesa Grande Rd.

RINCON SPRINGS
Observatory, 5 mi. E on State 76, then 8 mi. N on H-35.

SALTON CITY
Salton City Campground, 4 mi. E off State 86 via Marina Drive, on Salton Sea.

SANTA YSABEL
Lake Henshaw Resort, 7 mi. N via State 79, then 3½ mi. NW on State 76.

SIERRA CITY
Wild Plum, 2 mi. E via State 49 and forest road.

SONORA
Lyons Lake Resort, 18 mi. E on State 108.

TEHACHAPI
Tehachapi Mountain Park, 8 mi. SW off Highline Rd. via Water Canyon.

TEMECULA
Woodchuck, 9 mi. E via State 71, then 1 mi. S.

TWENTYNINE PALMS
Joshua Tree National Monument.

VICTORVILLE
Victorville KOA, 2½ mi. E off I-15 and U.S. 66.

WARNER SPRINGS
Indian Flats, 2½ mi. W on State 79, then 7 mi. N on S 05.

WILDROSE
Death Valley National Monument.

Colorado Campgrounds

ASPEN
Aspen Park, 3 mi. SE on State 82.

BAYFIELD
Pine Point E side of Vallecito Reservoir, reached from north off reservoir off U.S. 160.

BUENA VISTA
Cottonwood lake, 11 mi. W on forest road.

CANON CITY
Five Point Placer, 17 mi. W on U.S. 50.

CEDAREDGE
Crag Crest, on Forest Service Road 416, 3½ mi. W off State 65.

CREEDE
Lost trail, 38 mi. SW via State 149 and Rio Grande reservoir.

CUMBRES
Trujillo Meadows, off State 17, 3 mi. N of Cumbres pass.

DEL NORTE
Cathedral, 8 mi. W on U.S. 160, then 15 mi. N.

DELTA
Smokehouse, 5 mi. N of Delta-Nucla road from Columbine.

DOLORES
Burro Bridge, 37 mi. N on State 145 along Dolores River and left on W. Dolores road.

DURANGO
Columbine, 27 mi. N on U.S. 550.

EAGLE
Fulford Cave, 18 mi. SE on Brush Creek road.

FAIRPLAY
Buffalo Springs, 16 mi. S on U.S. 285, 2 mi. W.

FORT COLLINS
Ansel Watrous, on State 14, 23 mi. W of Ft. Collins.

GARFIELD
Garfield, off U.S. 50, 2 mi. above Garfield.

GRAND JUNCTION
Haypress, off U.S. 50 on county road 14 mi. S Glade Park store.

HAYDEN
Vaughn Lake, 21 mi. SE via state 317 and county roads 29 & 55.

LAKE CITY
Big Blue, 10 mi. W of State 140, 9 mi. N of Lake City.

LEADVILLE
Tennessee Pass, on U.S. 14, 10 mi. N of Leadville.

MARVINE
East Marvine, 4 mi. up Marvine Creek off Rio Blanco county rd.

MESA
Jumbo, 13 mi. S off State 65.

MONTE VISTA
Alamosa, off State 15, 25 mi. on Alamosa river road.

MONTROSE

Antone Springs, 25 mi. SW of Montrose off Divide Trail at jct.

OURAY

Amphitheatre, 1 mi. S of Ouray.

RED CLIFF

Blodgett, 4½ mi. S on U.S. 24.

REDSTONE

Bogan Flats, on Crystal River, 21 mi. S of Carbondale.

RIFLE

Little Box Canyon, 20 mi. NE off State 325 on fork of Little Rifle Creek.

SALIDA

North Fork Lakes, 12 mi. W off U.S. 50 on N. Fork South Ark. R.

SILVERTON

South Mineral 7 mi. N on U.S. 550 and forest roads.

STEAMBOAT SPRINGS

Box Canyon, 28 mi. N off U.S. 40 on Elk River.

TELLURIDE

Matterhorn, 10 mi. S on U.S. 550.

TOPONAS

Blacktail, on State 84, 11 mi. S of Toponas.

TRAPPER'S LAKE

Trapper's Lake, ¼ mi. from lake.

WHITEWATER

Carson Hole, 20 mi. SW on forest roads.

Idaho Campgrounds

BOISE
Alexander Flat, 70 mi. NE via Atlanta, State 21 and forest road.

CAREY
Copper Creek, 20 mi. N via forest road.

CHALLIS
East Fork, 18 mi. S on U.S. 93, at jct. of Main and East Fork Salmon Rivers.

CLAYTON
Holman Creek, 8 mi. W on U.S. 93.

COUNCIL
Bear Camp, 38 mi. NW via county and forest roads.

CROUCH
Hot Springs Campground, 6 mi. E via State 17.

FAIRFIELD
Pioneer, 10 mi. N on all weather road.

FREEDOM
Tincup, 5 mi. W on STATE 34.

GRANGERVILLE
Castle Creek Campground, 17 mi. SE on State 14.

HOLLISTER
Nat-Soo-Pah Campground, ½ mi. S on U.S. 93, then 3¾ mi. E.

KETCHUM
Baker Creek, 17 mi. N on U.S. 93.

LOWELL
Apgar Creek Campground, 14 mi. E on U.S. 12.

MACKAY
Mackay Reservoir, 4 mi. NW U.S. 93-A.

MOUNTAIN HOME
Big Roaming River Lake, 23 mi. NE via State 68, then 25 mi. N on forest road.

NORTH FORK
Ebenezer Bar, 8 mi. W via forest road.

ROGERSON
Bear Gulch, 15 mi. E.

SALMON
Cougar Point, 5 mi. S on U.S. 93, then 12 mi. W via forest road.

STANLEY
Alturas Lake, 20 mi. S via U.S. 93, then 4 mi. W on forest road.

TAMARACK
Cold Springs, 6 mi. W on forest road at res.

TWIN FALLS
Hunter's Trailer Park, E on U.S. 30.

WHITEBIRD
Skookumchuck, 4 mi. S on U.S. 95, on the Salmon River.

YELLOWPINE
Golden State Campground, 3 mi. S via graveled road.

Kansas Campgrounds

BENNINGTON
Ottawa County State Lake, 5 mi. N
and 1 mi. E.

BUFFALO
Wilson County State Lake, 2 mi. SE
on U.S. 75.

CARBONDALE
Osage County State Lake, 3 mi. S vis
U.S. 75, then 1 mi. E on U.S. 56,
then ½ mi. S.

CIMARRON
Cimarron Crossing campground, 8
blks. S of jct U.S.50 & State 23.

COUNCIL GROVE
Canning creek Cove, 2 mi. N on State
177, then 4 mi. W across dam on
county road.

JETMORE
Hodgeman County State Lake, 2 mi.
S on road along E city limits, then 4
mi. E.

MANKATO
Jewell County State Lake, 6 mi
S then 3 mi W on a county road.

MEDICINE LODGE
Barber County State Lake, N edge of
town.

OAKLEY
Camp Inn, ¼ mi. S from jct. with
1-70 on U.S. 83 & 383.

READING
Lyon County State Lake, 5 mi. W on
State 170, then 1½ mi. N.

RUSSELL
Waggoner Trailer Park, ½ mi. N of
jct. with Hwy 1-70 on U.S. 281.

RUSSELL SPRINGS
Logan County Lake, 1 mi. N and 2½
mi. W.

SALINA
J-D Campground, ¼ mi. N of jct
1-70, U.S. 40 & U.S. 81, then ½ mi.
W.

SCOTT CITY
Lake Scott State Park, 16 mi. N on
U.S. 83.

SYLVAN GROVE
Lucas Park, 6 mi. S on State 232.

Montana Campgrounds

ALDER
Alder KOA, ½ mi. E on State 287.

BIG TIMBER
Big Timber Campground, ¼ mi. SW on U.S. 10, at Boulder River Bridge.

BUTTE
Elk Park Campground, 16 mi. NE on U.S. 91.

CAMERON
Hilltop Campground, 22 mi. S on State 287

DEER LODGE
Phillipsburg Bay, 10 mi. S on U.S. 10 A, then 2 mi. S on W side of Georgetown Lake Forest Rd.

DILLON
Bannack State Monument, 20 mi. W off county 278.

ENNIS
Jack Creek Campground, 1 mi. S on State 287, then 12 mi. E on Jack Creek Road.

FORT PECK
Downstream, 1 mi. E Fort Peck Reservoir.

GALEN
Racetrack Campground, 12 mi. W on Racetrack Creek Road.

GLASGOW
Trails West Campground, ½ mi. W off U.S. 2.

GREAT FALLS
River Ranch, off U.S. 87 & 89, 6 mi. S on Fox Farm Road.

HARLOWTON
Deadman's Basin State Recreation Area, 20 mi. E off U.S. 12, on Deadman's Basin Reservoir.

HAVRE
Beaver Creek Park, From U.S. 2, on 5th Ave. then 10 mi. S on Beaver Creek Road.

LAKESIDE
West Shore State Park, 4 mi. S on U.S. 93.

LIBBY
Libby Creek, 12 mi. S on U.S. 2.

MALTA
Camp Creek, 52 mi. SW on U.S. 191, then 7 mi. W on county road and 1½ mi. N on access road.

NORRIS
Bear Trap, 6 mi. E on Secondary State 289.

SHERIDAN
Wright's Trailer Court, in town on U.S. 287.

SUPERIOR
Quartz Flat Campground, 11 mi. E on I-90.

THOMPSON FALLS
Fishtrap Creek Campground, 5 mi. E on U.S. 10-A, 15 mi. NE on Thompson River Road.

TROUT CREEK
Trout Creek Campground, 3 mi. W on U.S. 10-A.

TROY

Spar Lake Campground, 12 mi. S on State 202, then W on Spar Lake Road.

WHITE SULFUR SPRINGS

Aspen Campground, 44 mi. NE on U.S. 89.

Nebraska Campgrounds

AINSWORTH
Long Lake State Recreation Area, 8 mi. W on Hwy 20, then 26 mi. S on county road.

BRIDGEPORT
Bridgeport State Recreation Area, 1 mi. N of State 88.

CAMBRIDGE
Medicine Creek State Recreation Area, 9½ mi. NW.

CHADRON
Chadron State Park, 9 mi. S on U.S. 385.

CRAWFORD
Cochran State wayside Area, 5½ mi. S on State 2.

DUNLAP
Box Butte State Recreation Area, 5 mi. W on county road.

GERING
Wildcat Hills State Recreation Area, 10 mi. S on State 71.

HAYES CENTER
Hayes Center State Area, 12, mi. NE on dirt road.

HAYS SPRINGS
H & W Trailer Court, W side of town, at jct. U.S. 20 & State 87.

IMPERIAL
Enders Reservoir State Rec. Area, 8 mi. SE on U.S. 6 & State 61.

KIMBALL
Kimball KOA, ½ mi. E on U.S. 30.

LEWELLEN
Elm Court, in town, S of U.S. 26.

MC COOK
Karrer Park, on U.S. 6 & 34, E edge of town.

MERRIMAN
Cottonwood State Recreation Area, 1 mi. E on U.S. 20 & 1 mi. S on county road.

NORTH PLATTE
Cody Park, ½ mi. N on U.S. 83.

OGALLALA
Lake Mc Conaughy State Rec. Area, 9 mi. NE o n State 61.

RUSHVILLE
Smith Lake State Area, 23 mi. S on State 250.

SIDNEY
Trailer City, 1 mi. E on U.S. 30.

TRENTON
Swanson Reservoir State Recreation Area, 3 mi. W on U.S. 34.

VALENTINE
Big Alkali Lake State Area, 12 mi. S on U.S. 83, then 4 mi. W on State 483.

Nevada Campgrounds

CHURCHILL
Fort Churchill Historic State Monument, 8 mi. from crossroads at Silver Springs.

CURRANT
Currant Creek, 12 mi. SW on U.S. 6

DENIO
Sheldon National Antelope Range, 26 mi. W on State 8A.

ELKO
Wildhorse Reservoir, 66 mi. N on State 43.

ELY
Cleve Creek, 40 mi. SE on U.S. 6, 50 & 93, 10 mi. N of Spring Valley on county roads.

FALLON
Hub Totel, 3½ mi. W on U.S. 50 & 95.

HAWTHORNE
Alum Creek Campground, 8 mi. SW via State 31.

JACK CREEK
Jack Creek, 3 mi. NE.

JARBRIDGE
Pine Creek, 4 mi. S on forest road.

LAMOILLE
Thomas Canyon, 9 mi. SE on forest road.

LOVELOCK
Rye Patch Reservoir, 22 mi. N on 1-80, 1 mi. W to reservoir.

NORTH FORK
North Fork, 12 mi. N then 10 mi. SW.

OVERTON
Valley of Fire State Park, 6 mi. S on State 12, then W on State 40.

PARADISE VALLEY
Lye Creek, 16 mi. N on forest road.

RENO
Arrowhead Trailer Lodge, U.S. 395 to jct. of U.S. 40, then 2½ mi. W to 4175 U.S.

New Mexico Campgrounds

ABIQUIU
Echo Amphitheater, 16 mi. NW on U.S. 84.

AMELIA
Costilla River, on State 196 to Latir Lake.

BELEN
John F. Kennedy, 20 mi. E on county road.

CABALLO
Caballo Lake State Park, off U.S. 85 E.

CAPULIN
Capulin Mountain National Monument, 3½ mi. N of Capulin on State 325.

CONTINENTAL DIVIDE
Continental Divide Campground, on 1-40.

DEMING
City of rocks State Park, 22 mi. NW on U.S. 180, 5 mi. E on State 61.

GALLUP
Chaparral Overnight Mobile Inn, in town on U.S. 66.

GLENWOOD
Whitewater, 5½ mi. NE on forest road.

JEMEZ SPRINGS
Calaveras, 28 mi. N on State 126.

LORDSBURG
Yucca Trailer Court, W edge of town on U.S. 80 & 86.

MAGDALENA
Water Canyon, 12 mi. SE off U.S. 60.

MIAMI
Miami Lake, 8 mi. W on State 199.

MOGOLLON
Ben Lilly, E on State 78.

MONTICELLO
Springtime, 10 mi. N.

PECOS
Field Tract, 7 mi. N on State 63.

RADIUM SPRINGS
Lakewood Park, ½ mi. S on U.S. 85.

RATON
I.J. & N. Kampground KOA, 3 mi. S of Raton Pass on Business Route 85-87, on W side of hwy.

SAN LORENZO
Iron Creek, 15 mi. NE on State 90.

SILVER CITY
Scorpion Corral Campground, 50 mi. N on State 25.

WINGATE
McCaffey, 10 mi. S on State 400.

North Dakota Campgrounds

ASHELY
Lake Hoskins, 3 mi. W on State 3 & State 11.

BISMARK
Anderson Marina, 2¼ mi. W on U.S. 10, W end of the Memorial Bridge.

BURNSTAD
Beaver Lake State Recreation Area, 1 mi. N then 1 mi. E.

DICKINSON
Dickinson KOA, 1¼ mi. W on 195 business loop, then ¼ mi. S on State St.

JAMESTOWN
Museum Campground, 4 mi. E on 1-94, then S of 1-94 at Bloom Exit.

LINTON
Beaver Creek, 16 mi. W on gravel road.

MANDAN
Fort Lincoln State Park, 4½ mi. S on county 20.

NEW ENGLAND
Riverside Park, adjacent to town at jct. hwys, State 21 & 22.

NEW SALEMM
New Salem North Park, ½ mi. SW on State 31 and 1-94.

WATFORD CITY
Tobacco Garden Creek, 3 mi. E on State 34, then 29 mi. NE on county roads.

WERNER
Little Missouri, 2 mi. W and 10 mi. N on gravel road.

WILLISTON
Lewis & Clark, 18 mi. E then 3 mi. S via gravel road.

Oklahoma Campgrounds

ALTUS LAKE
18 mi. N of Altus.

ARDMORE
Ardmore KOA, 9 mi. S of Ardmore, off Oswalt road on 1-35

ATOKA
Atoka Lake, near Atoka.

BEAVER'S BEND
Beaver's Bend State Park, 11 mi. NE of Broken Bow on SH 259 and 259A on the Mountain Fork river.

BLACKWELL
Blackwell Lake, 12 mi. NW of Blackwell.

BROKEN BOW
Broken Bow Reservoir, 9 mi. N of Broken Bow on SH 259, then 1 mi. E.

CHECOTAH
City Lake, 4 mi. S of Checotah, then 1 mi. W.

CHEYENNE
Skip-Out Lake, 10 mi. W of Cheyenne on SH 41.

CHICKASHA
Chickasha Lake, 9 mi. NW of Chickasha.

CLINTON
Clinton Lake, 20 mi. W of Clinton.

DAVIS
Price's Falls, 2 mi. S of Davis on U.S. 77, then 4 mi. E on U.S. 77D.

DUNCAN
Clear Creek, NE of Duncan.

DURANT
Texoma Lake, Alberta Creek, SE of Kingston on SH 70.

FORT SUPPLY
Cottonwood Point, 3 mi. S of Fort Supply, off U.S. 183 & 270.

GUTHRIE
Guthrie Lake, 4 mi. S of Guthrie, then 1 mi. W.

HOBART
Hobart Lake, NW of Hobart.

LAWTON
Ellsworth Lake, 14 mi. N of Lawton.

PAULS VALLEY
Pauls Valley Lake, 2 mi. NE of Pauls Valley.

PAWHUSKA
Bluestem Lake, 6 mi. W of Pawhuska.

PAWNEE
City Lake, 1 mi. N of Pawnee on SH 18.

PERRY
Lake Perry, SW of Perry.

PONCA CITY
Ponca City Lake, 3 mi. E of Ponca Cityy.

RUSH SPRINGS
J.W. Taylor Lake, 3 mi, S of Rush Springs.

SEMINOLE
Sportsman Lake, 3 mi. E of Seminole on U.S. 270, then 2 miles N on county road.

SHAWNEE
Twin Lakes, 8 mi. W of Shawnee.

STILLWATER
Boomer Lake, N of Stillwater.

TAHLEQUAH
Tenkiller State Park, 7 mi. NE of Gore on SH 100.

WATONGA
Watonga Lake, near Watonga.

WAURIKA
Jap Beaver Lake, 4½ mi. NW of Waurika.

WOODWARD
Boiling Springs State Park, 3 mi. N of Woodward on SH 34, then 5 mi. E on SH 34C.

Oregon Campgrounds

ASHLAND
Emigrant Lake, 4 mi. SE on State 66.

BEND
Bend KOA, 3 mi. N on U.S. 97.

BURNS
Delintment Lake, 48 mi. NW off U.S. 20 on Donnelly Road and Delintment Lake Road.

CANYON CITY
Starr, 16 mi. S on U.S. 395.

CASCADIA
Cascadia State Park, off U.S. 20, W of town.

COOS BAY
Park Creek Recreation Area, 28 mi. E on Coos Bay Wagon Road to jct. with Middle Creek Access Road.

DALE
Big Creek, E off U.S. 395 on Ukiah-Granite Road.

DUFUR
Eightmile Crossing, 17 mi. W off U.S. 197 on Dufur Mill Road.

ENTERPRISE
Coyote, 45 mi. N off State 82 on Crow Creek Road.

ESTACADA
Armstrong, 17 mi. E on State 224.

FOSSIL
Shelton State Wayside, 10 mi. SE on State 19.

GLIDE
Big Twin Lakes, 40 mi. E off 1-5 on Little River Road 272 & Road 2715.

HARLAN
Big Elk, 1 mi. W off U.S. 20 on Big Elk Road.

HUNTINGTON
Farewell Bend State Park, 4 mi. SE on U.S. 30.

JOHN DAY
Canyon Creek Meadows, 20 mi. SE off U.S. 395 On Forest Road 1520.

JOSEPH
Blackhorse, 39 mi. SE off State 82 on Imnaha Road N. 111.

LA GRANDE
Hilgard Junction State park, 8 mi. W off 1-8 at jct. of Starkey Road.

MAUPIN
Shady Cove, 19 mi. NE off State 22 on Little North Salmon Road.

MADFORD
Fish Lake, 34 mi. NE off State 140 on Lake-Of-The-Woods Hwy.

MEHAMA
Shady Cove, 19 mi. NE off State 22 on Little North Salmon Road.

NEGANICUM
Saddle Mountain State Park, 8 mi. NE off U.S. 26.

NEWPORT
Beverly Beach Park, 7 mi. N on U.S. 101.

NORTH BEND
South Eel Creek, 15 N on U.S. 101.

PRAIRIE CITY
Dixie, 8 mo. NE on U.S. 26.

PRINEVILLE

Cougar, 25 mi. E on U.S. 26, 2 mi. W of Marks Creek Lodge.

SELMA

Deer Creek, 6 mi. E on Deer Creek county Road.

SHADY COVE

Shady Cove Trailer Lodge, S on State 62.

SPRAY

Bull Prairie, 17 mi. N off State 207 on Bull Prairie Rd.

STEAMBOAT

Apple Creek, 5 mi. SE on State 138.

SWEET HOME

Yellowbottom, 5 mi. E on U.S. 20, then NE 23 mi. on Quartzville Road.

UKIAH

Bear Wallow Creek, 10 mi. E on State 244.

South Dakota Campgrounds

BELLE FOURCHE
Herman Park, 1½ mi. S on 10th Ave. off U.S. 212.

CUSTER
Bismark Lake, 5 mi. E on U.S. 16-A.

DEADWOOD
Fish N Fry Campground, 5½ mi. S U.S. 385.

HERMOSA
Restway Travel Park, 6½ mi. W on State 36.

HILL CITY
Boarding House Gulch, 18 mi. N on U.S. 385.

HOT SPRINGS
Wind Cave National Park, Elk Mt. Campground, 10 mi. N of Hot Springs on U.S. 385.

INTERIOR
Cactus Flat Campground, ½ mi. S on U.S. 16A & State 40.

LEMMON
LLewellyn John's State Memorial, 11 mi. S on U.S. 73.

MURDO
Tee Pee Campground, just W from jct. with U.S. 83, on U.S. 16.

PINE RIDGE
Pine Ridge Campground, 1 mi. S on State 87.

RAPID CITY
Bradsky's Mobile Home Estates, E exit off 1-90, then 1¼ mi. to Campbell Ave. on State 79, ½ mi. from overpass.

ROSEBUD
Ghost Hawk Park, 4 mi. W of Rosebud.

SPEARFISH
Chris County Campground, 2 mi. S of Post Office on U.S. 14 & 85, then ½ mi. S on side road.

STAMFORD
Brave Bull, 3 mi. W on U.S. 16, then ½ mi. N on State 63.

WALL
Red Arrow Camp, in town, on U.S. 14, 16 & 1-90.

Texas Campgrounds

ALPINE
Roadrunner Park, 3¼ mi. SW on U.S. 67 & 90, then ¼ mi. N on gravel road.

AMARILLO
Amarillo KOA, 3½ mi. E on U.S. 60 & 66, then ¼ mi. S on side road.

BASTROP
Bastrop State Park, 2 mi. E on State 21.

BIG SPRING
City Park, in town, 2 mi. S on U.S. 87.

BROWNWOOD
Lake Brownwood State Park, 18 mi. NW off State 279 on Park Road.

BURNET
Inks Lake State Park, 13 mi. W on State 29.

CANYON
Palo Duro Canyon State Park, 12 mi. E on State 217.

COMMANCHE
Copperas Creek Park, 7 mi. E on U.S. 377, then 2½ mi. N on Road 2861.

CONCAN
G arner State Park, 7 mi. N on U.S. 83.

DALHART
Corral Trailer Park, ½ mi. E on U.S. 54.

FREDERICKSBURG
Enchanted Rock Park, 17½ mi. N on Ranch Road 965 from jct. with U.S. 290 & 87.

HUNTSVILLE
Huntsville State Park, 10 mi. S on U.S. 75 & 1-45.

KINGSLAND
Pogue's Fishing Camp, 1¼ mi. E on Ranch Road 1431, 5¼ mi. NE on Ranch Road 2342, then 1½ mi. NE on Park Road.

LEAKEY
Camp Flat Rock, ½ mi. E from With U.S. 83 on Ranch Rd. 1120.

MARBLE FALLS
Kemper's Korner, ½ mi. S on Main Street, then ¼ mi. W.

MATHIS
Lake Corpus Christi State Park, 6 mi. SW on State 359.

ROCKPORT
Goose Island State Park, 10 mi. N on U.S. 35, 1 mi. E on Park Road 13.

SNYDER
Trailertopia, 2½ mi. W on U.S. 180 from jct. with State 350.

SONORA
Caverns of Sonora, 8¾ mi. NW on U.S. 290 from jct. with U.S. 277, then 6½ mi. SW on Ranch Road 1989.

TOYAHVALE
Balmorhea State Park, ½ mi. NE on U.S. 290 from jct. with State 17.

UMBARGER
Buffalo Lake National Wildlife refuge, in town on Farn to Market Road.

UVALDE

Park Chalk Bluff, 16½ mi. NW on State 55 from jct. with U.S. 90 & 83, then 1¼ mi. S on a gravel road.

WASHINGTON

Washington-on-the-Brazos, 3 mi. SW off State 90.

ZAPATA

Falcon State Park, 30 mi. S on U.S. 83

Utah Campgrounds

BEAVER
Kent Lake, 16 mi. SE on State 153 & forest road.

BRIGHAM CITY
Willard Bay, 5 mi. S on U.S. 89, 91, & 30.

FILLMORE
Chalk Creek Trailer Park, N on U.S. 91.

GRANTSVILLE
South Willow Canyon Area, 11 mi. SW.

HANKSVILLE
McMillan Spring Campground, 33 mi. S via county road.

HATCH
Heart"O"the Parks, in town on U.S. 89.

KANAB
Little's Trailer Court, in town on U.S. 89, E of jct. U.S. 89 & 89A.

LEEDS
Oak Grove, 8 mi. NW on a forest road.

LOGAN
Friendship, 7 mi. S on State 163, then 7 mi. E on State 242, 4 mi. N on Saddle Creek.

MANTUA
Box Elder, adjacent to town, on U.S. 89.

MOAB
Dead Horse Point State Park, 12 mi. N on U.S. 160, then 22 mi. SW on State 279.

OAK CITY
Oak Creek, 8 mi. E on forest Road.

PAYSON
Payson Lake, 12 mi. SE on forest road.

PLEASANT GROVE
In town, on U.S. 91.

SAINT GEORGE
Bundy's KOA, 3 mi. E on 1-15.

TORREY
Capitol Reef, 12 mi. E of Torrey on State 24.

Washington Campgrounds

BARSTOW
Pierre Lake, 11 mi. N on Pierre Lake Road.

BURLINGTON
Bay View State Park, 14 mi. NW.

CARSON
Beaver, 18 mi. NW off U.S. 830 on Wind River Hwy.

CHEHALIS
Rainbow Falls State Park, 16 mi. W on State 6.

CLALLAM BAY
Lake Ozette Resort, 3 mi. W on State 112, then 20 mi. SW on Hoko-Ozette Road.

CLE ELUM
Beverly, 31 mi. N off U.S. 97 on North Fork Teanaway Rd.

COLVILLE
Twin Lakes Camp, 12 mi. E on State 294, then 6 mi. N on Twin Lakes Road.

CONCONULLY
Conconully State Park, N of town.

DAYTON
Edmiston, 21 mi. SE off U.S. 410 on Kendall-Skyline Road.

EASTON
Lake Easyon Resort, 1 mi. W off U.S. 10.

ELLENSBURY
Mineral Springs, 25 mi. NW on U.S. 97.

GOLDENDALE
Brooks Memorial State Park, 12 mi. N on U.S. 97.

JOYCE
Whiskey Creek Beach, 2½ mi. W on State 112, then ½ mi. N.

KETTLE FALLS
Lake Ellen Camp, 7 mi. W on State 30, 6 mi. S on Inchelium Road, then 3 mi. W on gravel road.

LA PUSH
The Surf Resort, 15 mi. W on U.S. 101.

MAZAMA
Ballard, 7 mi. NW off State 20 on Methow River Forest Hwy.

NACHES
House Creek on White Pass hwy near Tieton Ranger Station.

NEWPORT
Aqua Vista, on U.S. 2 & 195.

RANDLE
Adams Fork, 25 mi. SE off State 14 on Randle-Trout Lake Road.

SOAP LAKE
Patti-O-Park, 3 mi. N on State 17.

TWISP
Black Pine Lake, 11 mi. SW off State 20 on Poorman Creek Road.

WENATCHEE
Squillchuck State Park, 9 mi. S off county road.

WHITE SALMON
Bench Lake, 43 mi. N off U.S. 830 on Bird Creek Meadows Road.

WILLARD
Oklahoma, 9 mi. N off U.S. 830 on Little White Salmon County Road.

WINTHROP
Early Winters, 16 mi. NW on State 20, on Early Winters Creek.

YAKIMA
American Forks, 49 mi. NW off U.S. 410, at jct. Naches Hwy. & Bumping Lake Road.

Wyoming Campgrounds

BOULDER
Big Sandy, 60 mi. SE off U.S. 187.

CHEYENNE
Cheyenne KOA, 2½ mi. E on U.S. 30 & I-80.

CODY
Beartooth Lake, on U.S. 212, 24 mi. SE of Cooke City.

FORT WASHAKIE
Dickinson Creek, 25 mi. SW on forest roads.

GILLETTE
Tumble Weed Campground, 1 mi. W of jct. 59 & U.S. Business 16.

GLENDO
Esterbrook, 25 mi. W on forest roads.

GREYBULL
Camper's Court, 1 mi. N on U.S. 14, 1620 & State 789 at Twelfth St. N.

GUERNSEY
Guernsey State Park, 3 mi. W on county road, off U.S. 26.

JEFFREY CITY
Cottonwood, 6 mi. E on U.S. 287, then 10 mi. S on dirt road.

KEMMERER
Riverside Trailer Park, ½ mi. NE on U.S. 189, then ¼ mi. E on dirt road.

LOVELL
Bald Mountain, 37 mi. E on U.S. 14-A.

NEWCASTLE
Sikes Overnight Camping, 1½ mi. W on U.S. 16.

PINEDALE
Fremont Lake, 7 mi. NE off U.S. 87.

POWELL
Brownie's Trailer Court, ½ mi. SW of jct. Coulter Ave. & Bent St. on U.S. 14A.

RAWLINGS
Camp-A-Rama, ½ mi. NW on I-80 from jct. with U.S. 30 & 287, then ½ mi. N.

RIVERSIDE
Pelton Creek, 22 mi. SE on State 230.

SHERIDAN
Green Acres Campground, 1 mi. S of town, ½ mi. N of jct. U.S. 87 & 14.

SHOSHONI
Boysen State Park, 14 mi. N off U.S. 20.

THERMOPOLIS
Hot Springs State Park, in town.

WORLAND
Dunn's KOA, 1 mi. E on U.S. 16.

Arizona Motels

AJO
Marine Motel, N on State 85.

ALPINE
Country Club Motor Lodge, 4 mi. SE off U.S. 180.

BENSON
Sahara Motel, S on U.S. 80.

BULLHEAD CITY
Desert Aire Motel, 4½ mi. S on Needles Road.

CASA GRANDE
Francisco Grande Motor Inn, 5¼ mi W on State 84.

CASTLE HOT SPRINGS
Castle Hot Springs Hotel, in town.

DOUGLAS
Douglas Travelodge, E on U.S. 80, at 25-16th street.

GILA BEND
Desert Gem Motel, W on U.S. 80.

HOLEBROOK
City Center Motel, 615 W Hopi Dr.

KINGMAN
Holiday House Motel, 1¼ mi. NW on U.S. 93.

LAKE HAVASU CITY
Wings Motor Lodge, 33 Pima Drive.

MIAMI
Copper Hills Motor Hotel, 2½ mi E on U.S. 60.

NOGALES
Coronado Motel, N on U.S. 89, at 900 Grand Ave.

PAGE
Empire House Motel, 107 S 7th Ave.

PARKER
Desert Winds Motel, 700 Calif. Ave.

PAYSON
Diamond Dart Motel, N on State 87.

PHOENIX
Deserama Motel, 3 mi. E on U.S. 60, at 2853 E Van Buren.

SAFFORD
Sandia Motel, E on U.S. 70.

SALOME
Stanford Inn, E on U.S. 60.

SEDONA
Cedar Motel, S at jct. U.S. 89A & State 179.

SIERRA VISTA
Village Inn, 2 mi. E on State 90, at 2440 Fry Blvd.

SPRINGERVILLE
Reed's Motor Lodge, E on U.S. 60.

TUCSON
Imperial 400 Motel, 1284 N Stone Ave.

WICKENBURG
Rancho Grande Motor Hotel, 293 E Center St.

WILLCOX
Sands Motel, 400 S Haskell Ave.

WILLIAMS
Belaire Motel, W on U.S. 66.

WINSLOW
Desert Sun Motel, E on U.S. 66.

YUMA
Yuma Travelodge, 2050 4th Ave.

California Motels

AUBURN
Foothills Motel, 1½ mi. E off 1-80, exit via Auburn Ravine Road.

BAKER
Golden Choya Motel, on 1-15 & U.S. 91 business loop.

BAKERSFIELD
Hill House Motel, at 700 Truxtun Ave.

BARSTOW
Desert Inn Motel, ¼ mi. W of jct. 1-15 and U.S. 66 business route.

BISHOP
Elms Motel, 1 blk. E of U.S. 395 at 233 E Elm.

BORREGO SPRINGS
Stanlund's Resort Motel, 1 mi. S on county route S3 at 2771 Borrego Springs Road.

CALEXICO
Villa Sur Motel, 3 blks. E of SR 111, at 304 4th Street.

CHICO
Chico Travelodge, at 740 Broadway.

COLUMBIA
Columbia Gem Motel, 3 mi. N of Sonora, on Columbia Hwy.

CRESCENT CITY
Del Norte Motel, N edge of business route U.S. 101, at 975 9th Street.

DELANO
Stardust Motel, ¼ mi. N from SR 99 Freeway Southbound exit central Delano.

EL CENTRO
American Motel, at 725 State St.

EUREKA
The Fireside Motel, ½ mi. N on U.S. 101, at 1716 5th St.

FRESNO
The Carousel Motel, 1½ mi. N of Freeway 99 N & Belmont Ave. exit, at 1444 W White Ave.

GARBERVILLE
Motel Garberville, S edge of town, at 948 Redwood Drive.

GRASS VALLEY
Holiday Lodge Motel, on SR 20 & 49.

INDIO
El Morocco Motor Motel, 1 blk. S of U.S. 60, at 82-645 Miles Ave.

JACKSON
Amador Motel, 1½ mi. W in Martel, at SR 49 & 88 on Frontage Rd.

LAKE TAHOE
Blue Waters Motel, N off U.S. 50, at Cedar and Friday Streets.

LEE VINING
Gateway Motel, on U.S. 395.

LONE PINE
Frontier Motel, ½ mi. S on U.S. 395, at 1008 S Main Street.

MAMMOTH LAKES
Minaret Motel, on Minaret Summit Road.

MARYSVILLE
Imperial 400 Motel, on SR 20, at 721 10th Street.

MOJAVE
Imperial 400 Motel, 1 blk. E of SR 14, at 2145 Highway 58.

MORRO BAY
Bay View Motel, 222 Harbor Street, at Market Ave.

MOUNT SHASTA
Alpine Lodge Motel, at 908 S Mt. Shasta Blvd.

OROVILLE
The Villa Motel, at 1527 Feather River Blvd.

PALM SPRINGS
Tropics Motor Hotel, 1½ mi. S on SR 111, at 411 E Palm Canyon Drive.

PARADISE
Pink Lantern Motel, at 5799 Wildwood Lane.

PASO ROBLES
Cinderella Motel, 1¼ mi. N on U.S. 101 business route, at 2745 Spring Street.

PINE VALLEY
Hobart House Motel, on U.S. 80.

PLACERVILLE
Carter's Mother Lode Motel, 2 mi. E on U.S. 50, at 1940 E Broadway.

PORTERVILLE
Paul Bunyan Lodge, on SR 65, at 940 W Morton Street.

QUINCY
Spanish Creek Motel, NW edge on SR 70.

RED BLUFF
Cinderella Motel, 1 blk. E of city center, at 600 Rio Street.

REDDING
Bel Air Motel, 1 mi. N on 1-5 business loop.

SACRAMENTO
Arden Motel, at 1700 Del Paso Blvd.

SALTON CITY
Salton Bay Motor Hotel, 3 mi. E of SR 86.

SONORA
Modern Manor Motel, on SR 108, at 300 S Washington Street.

TRINIDAD
Bishop Pine Lodge, 2 mi. N on U.S. 101.

TWENTYNINE PALMS
Circle C Motel, 1½ mi. W & 2 blks. N of SR 62, at 6340 EL Ray Ave.

UKIAH
Royal Motel, on U.S. 101 business route, at 750 S State Street.

VISALIA

Cerrito's Marco Polo Motel, on SR 198, 1½ mi. W of jct. SR 63, at 4545 W Mineral King Ave.

WARNER SPRINGS

Warner's Resort, on SR 79.

YUBA CITY

Vada's Motel, on SR 20, at 545 Colusa Ave.

YUCCA VALLEY

Yucca Inn Motor Hotel, 1 mi. W on SR 62, 1 blk. N to 7500 Camino Del Cielo.

Colorado Motels

BOULDER
Boulder Travelodge, 1632 Broadway.

BUENA VISTA
Coronado Motel, N U.S. 24.

CANON CITY
Holiday Motel, E U.S. 50.

COLORADO SPRINGS
Colorado Springs Travelodge, 512 S Nevada Ave.

DILLON
Loveland Pass Motel, 7 mi. E U.S. 6.

FORT COLLINS
Edge O' Town Motel, 1½ mi. NW State 14.

GEORGETOWN
Georgetown Motor Inn, E U.S. 6.

GOLDEN
Holland House Motor Hotel, 1310 Washington Street.

GRAND JUNCTION
Bar X Motel, 2 mi. E on U.S. 6.

GREELEY
Greeley Travelodge, 721 13th Street.

GUNNISON
Western Motel, 411 E Tomichi Ave.

IDAHO SPRINGS
6 & 40 Motel, E U.S. 6.

KREMMLING
Bob's Western Motel, W U.S. 40.

LEADVILLE
Timberline Motel, 216 Harrison Ave.

MANITOU SPRINGS
Red Wing Motel, 56 El Paso Blvd.

MONTROSE
Red Arrow Motel, E U.S. 50.

OURAY
Antlers Motel, center U.S. 550.

Idaho Motels

ARCO
D – K Motel, E on U.S. 20.

BONNERS FERRY
Deep Creek Motel, 7 mi. S off U.S. 2.

BURLEY
East Park Motel, E on U.S. 30, at 507 E Main Street.

COEUR D'ALENE
Pines Motel, W on U.S. 10.

GRANGEVILLE
Elk Horn Motel, jct. U.S. 95 & E. Street.

KETCHUM
Tyrolean Lodge, SW off U.S. 93.

MC CALL
Park Motel, jct. State 55 & Lake Street.

MOUNTAIN HOME
Townhouse Motor Lodge, W on U.S. 20.

ST. MARIES
Pines Motel, 1117 Main Ave.

SALMON
Shady Nook Motel, N on U.S. 93.

SANDPOINT
Diamond W Motel, 4 mi. S on U.S. 95.

STANLEY
Redwood Motel, on U.S. 93.

TWIN FALLS
Dunes Motel, W on U.S. 30, at 447 Addison Ave.

WALLACE
Stardust Motel, W off U.S. 10, at 410 Pine Street.

Kansas Motels

BELLEVILLE
Plaza Motel, S on U.S. 36.

COLUMBUS
Townsman Motel, on U.S. 69, at 401 N East Ave.

CONCORDIA
Skyliner Motel, 2 mi. S U.S. 81.

FLORENCE
Holiday Motel, W U.S. 50.

GREAT BEND
Sands Motel, E U.S. 56, at 1015 E 10th Street.

HUTCHINSON
Astro Motel, 15 E 4th Street.

JUNCTION CITY
Park Terrace Motel, 2½ mi. E on 1-70.

KINGMAN
Kingman Motel, E U.S. 54.

MARYSVILLE
Thunderbird Motel, W U.S. 36.

MC PHERSON
Wheat State Motel, W U.S. 56.

MEDICINE LODGE
Copa Motel, S U.S. 160, at 401 W Fowler Street.

OAKLEY
Oakley Motel, N U.S. 83.

OSBORNE
Lazy L Motel, S U.S. 281.

RUSSELL
Kent Motel, 2 mi E U.S. 40.

SALINA
Salina Travelodge, on U.S. 81, at 245 S Broadway.

SCOTT CITY
Airliner Motel, E State 96.

WELLINGTON
Ramada Inn, 315 W 8th Street.

WICHITA
Branding Iron Motel, 5 mi. W on U.S. 54, at 6601 W Hwy. 54.

Montana Motels

BIG TIMBER
Lazy J Motel, E on U.S. 10.

BILLINGS
Billings Travelodge, 3311 2nd Ave.

BOZEMAN
Rainbow Motel, 510 N 7th Ave.

BUTTE
Capri Motel, 220 N Wyoming St.

COOKE CITY
Hoosier's Motel, W on U.S. 212.

DEER LODGE
Deer Lodge Motel, 819 Main St.

DILLON
Creston Motel, S on U.S. 91.

DRUMMOND
Star Motel, W off 1-90.

FORSYTH
Restwel Motel, center U.S. 10.

GARDINER
Town Motel, center U.S. 89.

GLENDIVE
El Centro Motel, 112 S Kendrick Ave.

GREAT FALLS
Sahara Motel, 3½ mi. SE on U.S. 87.

HARDIN
Lariat Motel, N on U.S. 87.

HARLOWTON
Corral Motel, E on U.S. 12.

HELENA
Holiday Motel, E on U.S. 12, at 1714 11th Ave.

LIBBY
Caboose Motel, W on U.S. 2.

LIVINGSTON
Parkway Motel, S on U.S. 89.

MILES CITY
Olive Motor Inn, off Main Street, at 15 N 5th Street.

SIDNEY
Park Plaza Motel, 115 6th St.

Nebraska Motels

ALLIANCE
Sunset Motel, E State 2.

ARAPAHOE
Mc Coy Motel, W U.S. 6.

CHADRON
Roundup Motel, E U.S. 20.

CRAWFORD
Hilltop Motel, S U.S. 20.

FREMONT
Berry's Motel, on U.S. at 445 W 23rd Street.

GERING
Circle S Lodge, on State 92, at 5th & M Streets.

HASTINGS
Grand Motel, 2 mi. S U.S. 6.

KIMBALL
Holiday Motor Lodge, 611 E 3rd Street.

LINCOLN
Buffalo Motel, 3 mi. E at 347 N 48th Street.

MC COOK
Cedar Motel, E U.S. 6.

NORTH PLATTE
Circle C Motel, 920 N Jeffers.

OGALLALA
Lazy K Motel, E U.S. 30.

RUSHVILLE
Antlers Motel, E U.S. 20.

SIDNEY
Deluxe Motel, W U.S. 30.

VALENTINE
Dunes Motel, SE U.S. 20.

Nevada Motels

BATTLE MOUNTAIN
Owl Motel, center U.S. 40.

BEATTY
Horseshoe Motel, S U.S. 95.

BOULDER CITY
Sands Motel, 809 Nevada Hwy.

CARSON CITY
Gateway Motel, 907 S Carson St.

ELY
White Pine Motel, E U.S. 50, at 1101 Aultman Street.

FALLON
Fallon Travelodge, 70 E Williams Ave.

FERNLEY
Starlite Motel, W U.S. 40.

HAWTHORNE
El Capitan Lodge, 540 F Street.

HENDERSON
Henderson Town House Motel, center U.S. 93, at 73 Water Street.

LOVELOCK
Sparky's Motel, E U.S. 40.

MC DERMITT
Mc Dermitt Motel, center U.S. 95.

SPARKS
Nugget Motor Lodge, 1225 B St.

TONOPAH
Sundown Motel, N U.S. 95.

VIRGINIA CITY
Sun Mountain Motel, S State 17.

WINNEMUCCA
Frontier Motel, E U.S. 40.

YERINGTON
Ranch House Motel, W Bridge Street.

New Mexico Motels

ALAMOGORODO
Alamogorodo Travelodge, 508 S White Sands Blvd.

ALBUQUERQUE
Crossroads Motel, 1½ mi. E on U.S. 66, at 1001 Central Ave.

CARLSBAD
Carlsbad Travelodge, E on U.S. 62, at 401 E Green Street.

CIMARRON
Kit Carson Inn Center, U.S. 64.

DEMING
Deming Travelodge, 500 W Pine Street.

FARMINGTON
Farmington Travelodge, W on U.S. 550, at 1510 W Main St.

GALLUP
Gallup Travelodge, 1½ mi. W on U.S. 66.

GRANTS
Grants Travelodge, 2 mi. E on U.S. 66.

HOBBS
Lamplighter Motel, S on U.S. 62, at 110 E Marland Street.

LAS CRUCES
Las Cruces Travelodge, 755 N Vallet Dr.

LAS VEGAS
Palomino Motel, N on U.S. 85.

LORDSBURG
Lordsburg Travelodge, E on U.S. 70.

RATON
Sands Manor Motel, S on U.S. 64.

ROSWELL
Roswell Travelodge, 2 mi. W on U.S. 70, at 2200 W 2nd.

SHIPROCK
Nataani Mez Lodge, center U.S. 666.

SILVER CITY
Drifter Motel, 2 mi E on U.S. 180.

North Dakota Motels

BISMARK
Bismark Motor Hotel, 2301 E Main Street.

BOWMAN
Trail Motel, W on U.S. 12.

CARRINGTON
Del Claire Motel, E on U.S. 52.

DEVILS LAKE
Artclare Motel, E on U.S. 2.

DICKINSON
Nodak Motel, E on 1-94.

FARGO
Midway Motel, 5 mi. W on U.S. 10.

GRAND FORKS
Plainsman Motel, 1½ mi. W on U.S. 2, at 2201 Gateway Drive.

HARVEY
R & R Motel, E on U.S. 52.

MINOT
Riverside Motor Lodge, 2 mi. E on U.S. 2.

RUGBY
Hamilton Motel, S on U.S. 2.

VALLEY CITY
Flickertail Motel, 2 mi. W on U.S. 10.

WILLISTON
Nite Cap Motel, 102 14th St.

Oklahoma Motels

ADA
Trails Motel, N on State 99.

CHICKASHA
Kings Inn Motel, 702 S 4th St.

CLAREMORE
Will Rogers Motor Court, S on U.S. 66.

DUNCAN
Century Motel, W on U.S. 81.

ELK CITY
Western Motel, 1½ mi. W on U.S. 66.

LAWTON
Lawtonian Hotel-Motel, 501 S 4th Street.

MIAMI
Thunderbird Motel, E on State 10.

MUSKOGEE
Trade Winds Motor Hotel, 1½ mi SW on U.S. 69, at 534 S 32nd.

NORMAN
Howard Johnson's Motor Lodge, 2 mi. W at jct. 1-35 & State 9.

OKLAHOMA CITY
Guest House Motel, 4 mi. N on U.S. 66, at 5200 N Classen Blvd.

PAULS VALLEY
4 Sands Motel, S on U.S. 77.

PERRY
Cherokee Strip Motel, 3 mi. W on U.S. 77.

PONCA CITY
Motel Thunderbird, on U.S. 77, at 407 S 14th Street.

SALLISAW
Stardust Motel, E on U.S. 64.

SHAWNEE
Holiday Harbor Motel, 10 mi. E on 1-40.

STILLWATER
Holiday Inn, 2 mi W on State 51.

Oregon Motels

ALBANY
Golden Door Motel, 2½ mi. E on U.S. 20, at 3125 Santiam Hwy.

ASHLAND
Motel Timbers, S on U.S. 99, at 1450 Ashland Street.

BAKER
Oregon Trail Motel, E on U.S. 30, at 211 Bridge Street.

BEND
Dunes Motel, N on U.S. 97, at 1515 3rd Street.

BIGGS
Riviera Motel, on U.S. 30 W of jct. 1-80 N.

BROOKINGS
Holiday Motel, N on U.S. 101.

BURNS
Ponderosa Motel, W on U.S. 20.

CANNON BEACH
Surfsand Resort Motel, on Oceanside & Division Street.

COOS BAY
Dunes Motel, N on U.S. 101, at 1445 N Bayshore Drive

CORVALLIS
Country Kitchen Motel, 800 N 9th Street.

DEPOE BAY
Surfside Motel, N on U.S. 101.

EUGENE
New Oregon Motel, S on U.S. 99, at 1655 Franklin Blvd.

FLORENCE
Silver Sands Motel, N on U.S. 101.

GOLD BEACH
Drift In Motel, on U.S. 101.

GRANT'S PASS
Grant's Pass Travelodge, 748 SE 7th Street.

HOOD RIVER
Meredith George Motel, 1¼ mi. W off 1-80 N at 8082 Westcliff Drive.

HUNTINGTON
Farewell Bend Motor Inn, 4 mi SE on U.S. 30.

KLAMATH FALLS
King Falls Motel, 2 mi. E on State 140, at 2660 Shasta Way.

LAKEVIEW
Rim Rock Motel, 727 S F St.

MADRAS
City Center Motel, center U.S. 97, at 124 5th Street.

NEWPORT
Jolly Knight Motel, S on U.S. 101, at 606 SW Coast Hwy.

NORTH BEND
Pony Village Motor Lodge W off U.S. 101 on Virginia Ave.

PRINEVILLE
Carolina Motel, E on U.S. 26.

ROSEBURG
Holiday Motel, 444 SE Oak St.

ST. HELENS
Village Inn, S on U.S. 30, at 535 S Hwy.

SHADY COVE
Royal Coachman Motel, center State 62.

SUTHERLIN
West Winds Motel, W at jct. 1-5 & State 138.

VIDA
Hawthorn Farm, on U.S. 126.

South Dakota Motels

BELLE FOURCHE
Lariat Motel, E on U.S. 212.

CUSTER
Chief Motel, W on U.S. 16, at 120 Mt. Rushmore Road.

DEADWOOD
Terrace Motel, E on U.S. 85.

EDGEMONT
Rainbow Motel, W on U.S. 18.

HILL CITY
Pine Edge Motel, 4½ mi. NE on U.S. 16.

HOT SPRINGS
Evans Heights Motel, N off U.S. 385.

KEYSTONE
Rushmore View Motel, S on U.S. 16A.

MARTIN
Harold's Motel, SE on U.S. 18.

RAPID CITY
Jensen's Motor Lodge, S on U.S. 16, at 1916 Mt. Rushmore Rd.

SPEARFISH
Siesta Motel, 2 mi. NW on U.S. 14.

STURGIS
Starlite Inn, S on State 79.

Texas Motels

ALPINE
Ponderosa Motor Inn, 2 mi. E on U.S. 90.

AMARILLO
Plainsman Motor Hotel, 2 mi. NE on 1-40, at 1530 E Amarillo Blvd.

AUSTIN
Howard Johnson's Motor Lodge, 6 mi. N on 1-35, at 7800 N Interregional Hwy.

BIG SPRING
Ponderosa Motel, S on U.S. 87.

CHILDRESS
Ranchouse Motel, W on U.S. 287.

DALLAS
Dallas Travelodge, 4001 Live Oak Street.

DEL RIO
Del Rio Travelodge, 1¼ mi. NW on U.S. 90.

EAGLE PASS
Colonial Inn Motel, 2 mi. NW on U.S. 277.

FORT WORTH
Fort Worth Travelodge, 4½ mi. S on 1-35 W at 3518 S Freeway.

FREDERICKSBURG
Sunday House Motel, E on U.S. 290, at 501 E Main Street.

GATESVILLE
Chateau Ville Motor Hotel, 1½ mi. E on U.S. 84, at 2501 E Main Street.

HEREFORD
Chateau Inn Motel, 500 W 1st.

HOUSTON
White House Motor Hotel, 5½ mi. SW on U.S. 90A, at 9300 S Main.

HUNTSVILLE
Holiday Inn, 1½ mi. W at jct. 1-45 & State 30.

KINGSVILLE
Holiday Inn, 1½ mi. E on U.S. 77.

LUBBOCK
Howard Johnson's Motor Lodge, 3¾ mi. S on U.S. 84, at 6011 Ave. H.

MIDLAND
Skyway Motel, 7 mi. W on U.S. 80.

MINERAL WELLS
12 Oaks Inn Motor Hotel, 2 mi. E on U.S. 180.

MONAHANS
Colonial Inn, 1¼ mi. S at jct. 1-20 & State 18.

ODESSA
Imperial 400 Motel, 221 W 2nd St.

PECOS
Pecos Travelodge, E on U.S. 80.

SONORA
Twin Oaks Motel, W on U.S. 290.

VAN HORN
Western Lodge, W on U.S. 80.

VERNON
Sands Motel, E on U.S. 287.

WACO
Sandman Motel, 3 mi. SW on State 6, at 3820 Franklin Ave.

WICHITA FALLS
Trade Winds Motor Hotel, Broad & Holiday Streets.

Utah Motels

BEAVER
Beaver Travelodge, 6th N. Main Street.

BLANDING
Gateway Motel, E State 47.

BLUFF
Recapture Court Motel, center State 47.

BRIGHAM CITY
Westward Ho Motel, 505 N Main.

CEDAR CITY
Cedar City Travelodge, 479 S Main Street.

DELTA
Delta Motel, 347 E Main St.

GREEN RIVER
Sleepy Hollow Motel, E on U.S. 6.

HATCH
New Bryce Motel, center U.S. 89.

HEBER
Wasatch Motel, 875 S Main St.

KANAB
Treasure Trail Motel, 146 W Center Street.

MILFORD
Park Motel, center off State 21.

MOAB
Apache Motel, 166 E 4th South.

MONTICELLO
Navajo Trail Motel, N U.S. 160.

ORDERVILLE
Fisher's Rancho Lodge, W U.S. 89.

PANGUITCH
Purple Sage Motel, 104 Center Street.

PROVO
Columbian Motel, S U.S. 89, at 70 East 300 South.

ST. GEORGE
Sands Motel, E U.S. 91.

SALINA
Pahvant Lodge, S U.S. 89.

Washington Motels

CHEHALIS
Cascade Motel, S off 1-5, at 13th St. & Kelly Rd.

CHELAN
Campbell's Lodge & Cottages, center U.S. 97.

CLARKSTON
Golden Key Motel, 1376 Bridge Street.

ELLENSBURG
Thunderbird Motel, 403 W 8th Ave.

ENUMCLAW
Harold's Motel, on State 169, at 548 Griffin Ave.

KALAMA
Columbian Inn Motel, N off 1-5.

KELSO
Motel 6, E off 1-5, at 1505 Kelso Drive.

NACHES
Game Ridge Motel, on U.S. 12, 20 mi. E of White Pass.

OROVILLE
Stone's Resort, W off U.S. 97, on Wannacut Lake.

PASCO
Holiday Motel, 720 W Lewis Street.

PROSSER
Prosser Motel, N on U.S. 12, at 6th Street.

PULLMAN
Pullman Travelodge, 405 S Grand Ave.

RAYMOND
Mountcastle Motel, W off U.S. 101, at 524 3rd Street.

WENATCHEE
Wenatchee Travelodge, 232 N Wenatchee Ave.

WESTPORT
Sportman Motel, S on Montesano Ave.

YAKIMA
Yakima Travelodge, 110 S Naches Ave.

Wyoming Motels

BUFFALO
Mountain View Motel, W on U.S. 16, at 585 Fort Street.

CASPER
Skyline Motel, 2 mi. SW on State 220, at 2037 Cy Ave.

CHEYENNE
Cheyene Travelodge, W on U.S. 30, at 1100 W 16th Street.

DOUGLAS
Chieftain Motel, 1½ mi. E on U.S. 20.

DUBOIS
Branding Iron Motel, W on U.S. 26.

GREEN RIVER
Coachman Inn, E on U.S. 30, at 470 E Flaming Gorge Way.

GREYBULL
Yellowstone Motel, E on U.S. 14.

JACKSON
Western Motel, 101 W Simpson Street.

KEMMERER
Antler Motel, 419 Coral St.

LOVELL
Horseshoe Bend Motel, 375 E Main Street.

POWELL
King's Motel, E on U.S. 14A.

RAWLINGS
West Way Lodge, W on U.S. 30, at 1219 W Spruce Street.

RIVERTON
Tomahawk Motel, 208 E Main Street.

ROCK SPRINGS
El Rancho Motor Lodge, E on I-80, at 1430 9th Street.

THERMOPOLIS
California Motel, 501 S 6th.

TORRINGTON
Western Motel, 1½ mi. NW on U.S. 26.

WHEATLAND
Wyoming Motel, N on U.S. 87.

Bibliography

BOOKS

Dake, H.C. *California Gem Trails*. California: Gemac Corp.

Duke, A. *Arizona Gem Fields*. California: Gemac Corp.

Johnson, H.C. *Western Gem Hunters Atlas*. California: Cy Johnson, 1971.

Mac Fall, R.P. *Gem Hunters Guide*. New York: Crewell Co., 1963.

Pearl, R.M. *Colorado Gem Trails & Mineral Guide*. Sage Books. 1958.

Pearl, R.M. *How To Know The Minerals & Rocks*. New York: McGraw-Hill, 1955.

Ranson, J.E. *Rock Hunters Range Guide*. New York: Harper, 1962.

Sanborn, W.B. *Crystal & Mineral Collecting*. California: Gembooks, 1966.

Shaffer, H.S. Zim & P.R. *Rocks & Minerals*. New York: Golden Press, 1957.

Simpson, B.W. *Gem Trails of Texas*. Gem Trails, 1958.

Simpson, B.W. *New Mexico Gem Trails*. California: Gemac Corp.

MAGAZINES

Desert Magazine, Palm Desert, Calif.

Earth Science, Box 1357, Chicago, Ill.

Gems & Minerals, Box 687, Mentone, Calif.

Lapidary Journal, P.O. Box 518, Del Mar, Calif.

Rocks & Minerals, Box 29, Peekskill, New York